Comparative studies on the pharyngeal teeth of cyprinids

Comparative studies
on the pharyngeal teeth of cyprinids

Tsuneo Nakajima

Tokai University Press

Tsuneo Nakajima

Professor, Okayama University of Science, Okayama 2012 to present
Curator, Lake Biwa Museum, Kusatsu 1991 to 2010
Researcher, Gifu College of Dentistry, Gifu 1980 to 1991
Winning Biwako Prize for Ecology, 1997
PhD in Science, Kyoto University, Kyoto 1980

Comparative studies on the pharyngeal teeth of cyprinids
Tsuneo Nakajima

©2018 Tsuneo Nakajima
All right reserved, but the rights to each figure belong to the person stated in the text.
No part of this book may be reproduced in any form or any other means
without the written permission of the publisher.

ISBN978-4-486-03738-5
First published January 2018
Printed in Japan
Tokai University Press
4-1-1, Kitakaname, Hiratsuka-shi, Kanagawa 259-1292, Japan

Contents

Preface   vii

1. Introduction ·· 1

1-1. Larval dentition and adult dentition ·· 1

1-2. Tooth replacement pattern ·· 2

1-3. Variation in the larval dentition ·· 3

1-4. Formation of minor rows ·· 6

1-5. Differences between cyprinid and cobitid dentitions ············· 6

2. Morphogenesis of pharyngeal teeth in some typical cyprinid species ·············· 9

2-1. Morphogenesis of pharyngeal teeth in *Carassius auratus grandoculis* ·················· 9

   Developmental changes in the dentition of larvae and juveniles ·················· 9

   Developmental stages of teeth in *Carassius auratus grandoculis* ·················· 11

2-2. Morphogenesis of pharyngeal teeth in *Acrossocheilus parallens* ·················· 11

   Developmental changes in the larval dentition or A-row teeth ·················· 11

   Developmental stages of teeth in *Acrossocheilus parallens* ·················· 13

2-3. Morphogenesis of pharyngeal teeth in *Gnathopogon elongatus* ·················· 14

2-4. Common stages of tooth development ·· 15

2-5. Developmental changes of the central teeth in some cyprinids ·················· 15

   2-5-1. *Opsariichthys bidens* ·· 15

   2-5-2. *Spinibarbus sinensis* ·· 15

   2-5-3. *Megalobrama amblycephala* ·· 17

   2-5-4. *Hypophthalmichthys nobilis* ·· 17

   2-5-5. *Rhodeus ocellatus* and *Acheilognathus rhombeus* ·················· 18

   2-5-6. *Ctenopharyngodon idella* ·· 20

   2-5-7. *Tribolodon hakonensis* ·· 20

   2-5-8. *Cirrhinus molitorella* ·· 20

   2-5-9. *Pseudorasbora parva* ·· 21

   2-5-10. *Sarcocheilichthys sinensis* ·· 21

2-5-11. *Tinca tinca* ·· 21

2-5-12. *Mylopharyngodon piceus* ··· 21

2-5-13. *Cyprinus carpio* ·· 21

2-6. Developmental stages in cyprinid teeth ····························· 21

3. Descriptions of the pharyngeal dentition of cyprinid subfamilies ················· 27

Dental formula ·· 27

Morphorogical types based on developmental stages of teeth ············· 27

3-1. Danioninae ·· 29

3-2. Barbinae ·· 45

3-3. Schizothoracinae ·· 73

3-4. Cultrinae ··· 80

3-5. Xenocypridinae ··· 89

3-6. Hypophthalmichthyinae ·· 91

3-7. Achielognathinae ··· 92

3-8. Leuciscinae ·· 100

3-9. Labeoninae ·· 113

3-10. Gobiobotinae ·· 123

3-11. Gobioninae ·· 126

3-12. Cyprininae. ·· 139

4. Summary ··· 147

4-1. Developmental stages and types of pharyngeal teeth ·················· 147

4-2. Dentition of each subfamily ··· 149

4-3. Nomenclature of pharyneal dentition ·· 151

Terminology for pharyngeal dentitions ·· 152

Terminology for tooth parts ··· 153

References   155

Postscript   157

Index   159

Preface

During vertebrate evolution, the specialized diphyodont, oligodont, plexodont, and heterodont dentitions of various mammals were derived from polyphyodont, polydont, haplodont, and homodont ancestral situations, which are still found in many living vertebrates, including fish, amphibians, and reptiles. Although the dentition of cyprinid fish is fundamentally polyphyodont, specialized oligodont, plexodont, and heterodont dentitions also occur in the Cyprinidae, uniquely among fish.

Bony fishes usually have five pairs of branchial arches, four of which bear gill rakers medially and gill lamellae laterally. Some have the fifth arch more or less modified into a pharyngeal bone bearing teeth. Cyprinid fishes, lacking true teeth in the oral cavity, have a well-developed dentition on this pair of gill-arch-derived pharyngeal bones. The presence and arrangement of these pharyngeal teeth is one of the important diagnostic characters of the family Cyprinidae (Berg, 1940). All cypriniform fish lack teeth on their jaws and in their oral cavity while possessing pharyngeal teeth, but only cyprinids among the Cypriniformes establish a fixed arrangement and number of pharyngeal teeth by the beginning of the juvenile period (Nakajima, 1987). The shape, arrangement, and number of these teeth provide very important clues for classifying the species (Regan, 1912; Chu, 1935).

Because pharyngeal teeth are composed of the hardest tissue of any found in fish and are very resistant to decay, they are frequently well preserved as remains in geological strata and archaeological sites. The study of pharyngeal teeth of cyprinid fishes can be very revealing. From the characteristics of the teeth we can tell much about the fish themselves and their living conditions. We can also follow the evolution of cyprinids (e.g., Vasnecov, 1939; Nakajima, 1998), geohistorical changes in freshwater systems (e.g., Nakajima and Yamasaki, 1992; Nakajima, 1994, 2012), and also human activities involving the fish (e.g., Nakajima, 2006; Nakajima et al., 2010).

Cyprinids are distributed mainly on the Eurasian continent, in North America and Africa, and in these continents' large peripheral islands. They comprise the largest freshwater fish family, with approximately 200 genera and 2,500 species (Nelson, 2016). They bear various kinds of pharyngeal dentition. Chu (1935) described the pharyngeal dentition of many Chinese cyprinids, but no publication has treated the pharyngeal dentition of the cyprinids as a whole. Furthermore, only a few studies besides that of Vasnecov (1939) were made on developmental changes in pharyngeal dentition until the end of the 1970s (Nakajima, 1979). Since then, several authors have described the development of the dentition in certain cyprinids (Nakajima, 1984; He et al., 1994; Huysseune et al., 1998; Van der Heyden and Huysseune, 2000; Van der Heyden et al., 2001, Shan, 2001), and morphogenesis of the pharyngeal teeth has been studied using SEM (Kodera, 1982; Nakajima, 1990; Nakajima and Yue, 1989, 1995; Yue and Nakajima, 1994; Sato et al., 2000).

The purpose of this volume is to offer SEM images of the pharyngeal dentitions of a wide range of cyprinids for the first time, based on materials preserved mainly at the Lake Biwa Museum in Japan and the Institute of Hydrobiology, Chinese Academy of Sciences, and to show how the pharyngeal dentition is expressed in various groups (subfamilies) of cyprinids, but not to describe differences among individual species. To accomplish this, the morphogenesis of the pharyngeal teeth of several cyprinids is described for the first time. Based on these data and earlier descriptions, a new categorization of ontogenetic patterns of these teeth is proposed, which serves as the basis for describing the teeth of a wide range of cyprinids representing every subfamily.

1. Introduction

1-1. Larval dentition and adult dentition

Nakajima (1979, 1984, 1990) studied the development of the dentition in several cyprinids. Although the adult fish bear a diverse assortment of pharyngeal dentitions, the larvae are similar to each other in dentition arrangement and tooth shape. The pharyngeal teeth of larvae are multi-rowed, and the initial teeth are recurved and conical (Fig. 1-1), quite different from the adult state. Nakajima (1979) called the multi-row condition of larvae the "larval dentition." The period during which the larval dentition persists does not necessarily correspond to the larval phase of fish development.

The definitions of the larval and adult dentitions give below and the terms applied to both dentitions are based on earlier descriptions of the dentition of several species of cyprinid larvae and juveniles (Nakajima, 1979; 1984, 1990). In the descriptions that follow, a "replacement wave" is a set of replacement teeth all of the same generation. The teeth appear at alternate positions ("even" or "odd") from posterior to anterior along the pharyngeal bone. The teeth of a single replacement wave ankylose to the bone at either even or odd positions, so the waves are correspondingly called even or odd replacement waves. The initial tooth (progenitor) and the replacement teeth that appear in sequence at each tooth position comprise a "tooth family" (Fig. 1-2).

In general, the pharyngeal teeth occur in one to three rows in adult cyprinids (Vladykov, 1934; Chu, 1935; Eastman and Underhill, 1973), and rarely in four rows (Shkil et al., 2010). The tooth rows are called the "A row", "B row", and "C row" from the medial side of the pharyngeal bone lateralwards in multi-row dentitions. Because the teeth of the A row are larger than those of other rows, the A row is sometimes also called the "major row," and other rows "minor rows." The tooth positions are commonly numbered from anterior to posterior in each row. For example, position A1 is the most anterior position of the A row. Tooth A1 is the all-inclusive

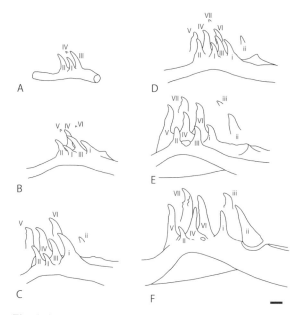

Fig. 1-1. Developmental changes in right larval dentition of *Gnathopogon caerulescens*, showing both ankylosed teeth and non-ankylosed tooth germs. A, 5.2 mm SL; B, 7.2 mm SL; C, 8.8 mm SL; D, 9.9 mm SL; E, 10.9 mm SL; F, 11.9 mm SL. I-VII, central teeth; i-iii, anterior teeth. Scale shows 10 μm, (from Nakajima, 1979)

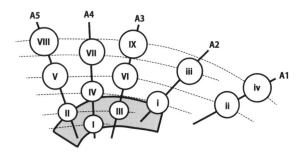

Fig. 1-2. Tooth distribution and order of appearance in right larval dentition of *Gnathopogon caerulescens*. Shaded area shows the initial state of the pharyngeal bone. Solid lines show tooth families, and broken lines show replacement waves. (from Nakajima, 1979)

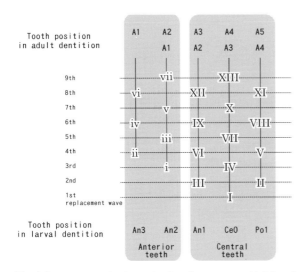
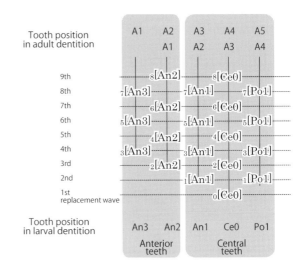

Fig. 1-3. Tooth distribution and order of appearance of left larval dentition: general scheme for cyprinids. Solid lines show tooth families and broken lines show replacement waves.

Fig. 1-4. General scheme for assigning tooth numbers to larval dentition in cyprinids and the relation between tooth positions in larval and adult dentitions.

term for the teeth successively cut at position A1 and belonging to the same tooth family. The adult dentition, which is represented by a dental formula commonly utilized in systematic descriptions (see below), is seen in both adults and juveniles. In the adult dentition, the major and minor rows have all been formed, and a tooth row consists of teeth originating in multiple replacement waves.

Some of the teeth ankylose to the curved part, and others to the anterior part of the larval pharyngeal bone. The latter are called the "anterior teeth", which are numbered in order of appearance with small Roman numerals (i, ii, iii, etc.). The former are called "central teeth," which are numbered in order of appearance with capital Roman numerals (I, II, III, etc.) (Fig. 1-3). The position where the initial tooth (I) appears is numbered Ce0; more anterior tooth positions are numbered successively An1, An2, An3, An4, and more posterior ones are numbered successively Po1, Po2, etc. The general formula for position (r) and replacement wave number (n) is $_{n-1}[r]$; thus the tooth position Ce0 in the first wave, in which the initial tooth (I) arises, would be identified as $_0[Ce0]$ (Fig. 1-4).

In adults and juveniles, the functioning tooth at any given position is generally shed when the replacement tooth has ankylosed to the bone or when the germ of the replacement tooth has grown sufficiently. In the larval dentition, however, the replacement tooth ankyloses medial to the functioning tooth, so the older functioning tooth is retained after ankylosis of the replacement tooth. Therefore, tooth rows of replacement teeth are formed medially of the original teeth, and the dentition becomes multi-rowed. The larval dentition is defined as a multi-row dentition in which the teeth of each row belong to a single replacement wave. Each tooth row in the larval dentition thus consists of teeth ankylosed alternately at even and odd positions.

In fish that have a multi-row adult dentition, some tooth germs appear lateral to the functioning teeth during the later post-larval to early juvenile period. These germs develop into the teeth of the minor row in the adult dentition. During the period between the appearance of the first minor-row tooth and the completion of the adult dentition, a "transitional dentition" is present. It can be difficult to identify each lateral tooth in the transitional dentition.

1-2. Tooth replacement pattern

Cyprinid dentition is polyphyodont and the fish continue to replace their pharyngeal teeth in a regular pattern throughout their life span. The replacement waves sweep alternately and cephalad in common cyprinids (Evans and Deubler, 1955; Nakajima, 1979; Nakajima et al., 1981, 1983), as is also known in many reptiles (Edmund, 1960). The pharyngeal dentition undergoes changes in both tooth shape and size in the course of these tooth replacements.

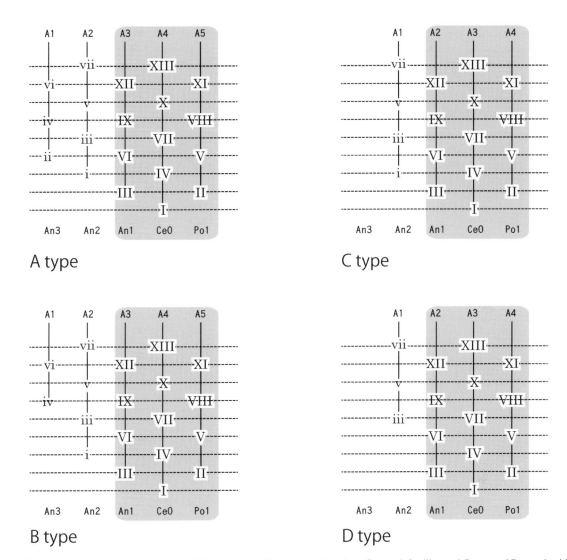

Fig. 1-5. Four types of larval dentition in cyprinids. A-type and B-type dentitions have five tooth families, and C-type and D-type dentitions have four tooth families. All have three families of central teeth (grey area) but they differ in the number of anterior tooth families. Anterior tooth ii or i fails to appear in the early replacement waves in B-type and D-type dentitions respectively, which are thus "imperfect" larval dentitions. In contrast, A-type and C-type dentitions are "perfect" larval dentitions. (from Nakajima, 1984)

Several models have been put forward to explain the jaw-tooth replacement mechanism in polyphyodont vertebrates (Edmund, 1960, 1962; Osborn, 1971, 1972, 1974, 1978; Lawson et al., 1971; DeMar, 1972). Of these, the "Zahnreihe" model of Edmund (1962) first provided a common basis for the comparative study of replacement patterns. To explain the regular patterns of tooth replacement, he postulated hypothetical stimuli that initiate tooth development. Osborn (1971) studied the embryonic development of the pattern of tooth replacement and concluded that his findings were not consistent with the "Zahnreihe" model. He constructed a different model, based on the view that the patterns are self-generated rather than extrinsically controlled (Osborn, 1971, 1972, 1974), and extended this self-generating model to account for development of gradients of tooth shape (Osborn, 1978). The appearance pattern of tooth germs in cyprinid larvae is consistent with Osborn's (1971) model (Nakajima, 1979, 1990).

1-3. Variation in the larval dentition

Although larval dentitions are similar to each other among cyprinid species, they vary with respect to the anterior teeth. Larval teeth become ankylosed both to the curved part (central teeth) and anterior elongated

Table 1-1 Types of larval dentition and dental fomulae of the adult dentition (after Nakajima, 1984)

Species	Types of larval dentition (frequency: %)	Number of tooth families	Dental formulae of adult dentition	Number of teeth in adult dentition
Danioninae				
Zacco platypus	A, B, C, or D	4 or 5	1(2).3(4).4(5)‒4(5).3.1(2)	4(5)‒4(5)
Opsariichthys uncirostris	A, B, C, or D	4 or 5	1(2).3(4).4(5)‒4(5).3.1(2)	4(5)‒4(5)
Leuciscinae				
Ctenopharyngodon idella	B‒D (50)	5‒4	2.4(5)‒4(5).2	4(5)‒4(5)
	D‒B (33.3)	4‒5		
	D‒C (16.7)	4‒4		
Tribolodon hakonensis	B‒D (79.0)	5‒4	2.5‒4.2	5‒4
	C‒D (16.7)	4‒4		
	D‒D (4.3)	4‒4		
Gobioninae				
Gnathopogon caerulescens	A‒A (100)	5‒5	2(3,4).5‒5.2(3,4)	5‒5
Gnathopogon elongatus	A‒A (100)	5‒5	2(3).5‒5.2(3)	5‒5
Pseudogobio esocinus	A‒A (100)	5‒5	2(3,4,5).5‒5.2(3,4)	5‒5
Sarcocheilichthys variegatus microoculus	A‒A (86.7)	5‒5	5‒5	5‒5
	B‒A (33.3)	5‒5		
Pungtungia herzi	A‒A (100)	5‒5	5‒5	5‒5
Pseudorasbora parva	A‒A (100)	5‒5	5‒5	5‒5
Hemibarbus barbus	B‒B (66.7)	5‒5	1.3.5‒5.3.1	5‒5
	A‒B (33.3)	5‒5		
Cypriniae				
Cyprinus carpio	C‒C (100)	4‒4	1.1.3‒3.1.1	3‒3
Carassius auratus grandoculis	C‒C (100)	4‒4	4‒4	4‒4

part (anterior teeth) of the initial pharyngeal bone (see Figs. 1‒1 to 1‒4). The central teeth appear at three positions: the position of the initial tooth (position Ce0), and the positions immediately posterior and anterior to it (positions Po1 and An1). All the central teeth of the larval dentition appear in every cyprinid species, but the appearance pattern of the anterior teeth, which appear at positions An2 and An3, varies between and within

different species. The types of larval dentition can be classified into four types, A type, B type, C type, and D type (Fig. 1‒5: Nakajima, 1984), with respect to the appearance pattern of the anterior teeth. A- and B-type larval dentitions have five tooth families, and C- and D-type larval dentitions have four tooth families. The first tooth at position An3 in the fourth replacement wave (tooth ii: $_3$[An3]) does not appear in the B-type

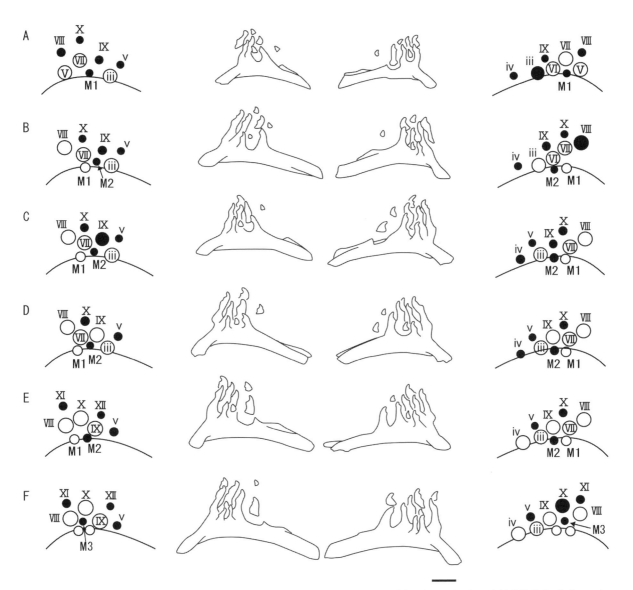

Fig. 1-6. Formation of the minor row of the pharyngeal teeth in successively order individuals (A-F) of cyprinid *Tribolodon hakonensis*. Note that the left and right sides differ. V-XII, central teeth; iii-v, anterior teeth; M1-M3, minor-row teeth. Scale shows 200μm. (from Nakajiama, 1990)

dentition, and the first tooth at position An2 in the third replacement wave (tooth i: $_2$[An2]) does not appear in the D-type dentition. These may be termed "imperfect type of larval dentitions," in which certain anterior teeth are non-existent in the early replacement waves. In contrast, A- and C-types are "perfect type of larval dentitions."

Since Nakajima (1984) proposed this classification, additional types have been observed by some authors. Husysseune *et al.* (1998) and Van der Hyden and Huysseune (2000) reported a variation of the B-type larval dentition in zebrafish, *Danio rerio*. The first tooth at position An2 (tooth i: $_2$[An2]) and the first and second teeth at An3 (teeth ii and iv: $_3$[An3] and $_5$[An3]) do not appear, and this species has an imperfect type of larval dentition involving five tooth families (see Figs. 1-3 and 1-4). Shan (2001) described the development of the dentition in *Distoechdon compressus*, which has seven teeth in the major row. According to her description, there are several types of larval dentition, all imperfect.

Since the A- and B-type larval dentitions have five tooth families and the C- and D-type larval dentitions have four tooth families (Nakajima, 1984), the A-type and B-type larval dentitions both develop into an adult dentition with five teeth in the major row, while the C- and D-type larval dentitions both develop into an adult dentition with four teeth in the major row. The types of larval dentition and number of major-row teeth in several cyprinid species are shown in Table 1-1.

Although common carp, *Cyprinus carpio*, has only three teeth in the major row, its larval dentition is of the D type with four tooth families. The number of tooth families is not necessarily consistent with the number of major-row teeth. Common carp keep four major-row teeth up into the early juvenile period. Subsequently, one tooth, which had been ankylosed at position Po1, is shed without the appearance of a replacement tooth, and three teeth remain in the major row thereafter (Kodera, 1982).

The teeth of the larval dentition develop into teeth of the A row of the adult dentition through tooth replacement. The tooth positions of the adult dentition are numbered from anterior to posterior in each row, and the tooth positions of the larval dentition are numbered on the basis of their situation relative to the initial tooth $_1$[Ce0]. To reconcile these distinct nomenclatures, I have marshalled data about tooth positions in the larval dentitions and the A row in the adult dentitions for a number cyprinid species in an easy-to-understand manner (Table 1-1). Briefly, the second position from the back in the A row of the adult dentition corresponds to position Ce0 of the larval dentition in all cyprinids except for *Cyprinus carpio*, in which, owing to the ontogenetic loss of the most posterior tooth as described above, the most posterior position of the A row corresponds to position Ce0.

1-4. Formation of minor rows

Nakajima (1990) described the formation of the minor row in the Japanese dace, *Tribolodon hakonensis*. The tooth germ of the first minor-row tooth appears lateral to all teeth of the larval dentition in the late post-larval period, and it ankyloses to the bone at the end of the post-larval period. The minor row is fully formed by the beginning of the juvenile period, whereby the adult dention is completed. The tooth germs of the minor row appear independently and evidently have a different developmental pattern from those of the major row (Fig. 1-6).

He *et al.* (1994) have observed the larval dentition of *Gobiocypris rarus* and described a different formation of the minor row. According to their description, the old lateral larval teeth at positions Po1 and Ce0 are not shed; instead, the tooth germs of their replacement teeth appear and form the minor row.

The minor rows appear later than the major rows in ontogeny. They first appear during the latest post-larval period and persist in the juvenile and adult. The major row develops in two phases. The change from the multi-row larval dentition of the first phase into the adult major row during the second phase is precipitated by the appearance of the tooth germs of the minor rows. Why should the appearance of the minor rows have such an effect on the pre-existing teeth?

Nakajima (1987) hypothesized that the larval dentition closely resembles that of the ancestral form of Cypriniformes. The multiple rows of the larval dentition develop as the pharyngeal bone broadens, thereby providing a broad dentigerous surface that makes ankylosis of the minor row possible. The lateral position of this first minor row is, however, perhaps unexpected. The teeth of the larval dentition are arranged in order of their appearance from lateral to medial positions, the lateral teeth being the older and the medial teeth the younger of the larval rows. The minor row that develops during the latest post-larval period is lateral to the major row of the adult dentition, yet it is younger than the most medial row of the larval dention.

1-5. Differences between cyprinid and cobitid dentitions

Nakajima (1987) described the development of the cobitid dentition in detail. One tooth ankyloses at position Ce0, and two tooth germs of the second replacement wave appear at positions Po1 and An1. These two teeth then ankylose at positions Po1 and An1, after which these three teeth all simultaneously ankylose to the bone as in the cyprinid laraval dentition. The teeth in cobitids increase in number with development, in the following manner. When the teeth of an even replacement wave ankylose, the bigger tooth germs in the following odd replacement wave arise medial to the ankylosed teeth. The smaller tooth germs in the following even replacement wave arise medial to these, and their number is also one more than that of the ankylosed teeth. On the other hand, when the teeth of an odd replacement wave ankylose, the bigger tooth germs on the following even replacement wave arise medial to the ankylosed teeth, with no change in number. The smaller tooth germs in the following odd replacement wave arise medial to the bigger tooth germs, with an increase in number by one compared to the ankylosed teeth. The cobitid dentition

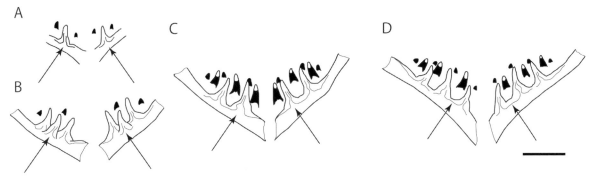

Fig. 1-7. Developmental changes in pharyngeal dentition of larval cobitid *Misgurnus anguillicaudatus*. Arrows show position Ce0. Developing tooth germs are shown in black. A, 50 mm SL; B, 6.1 mm SL; C, 6.9 mm SL; D, 7.9 mm SL. Scale shows 100μm. (from Nakajima, 1987)

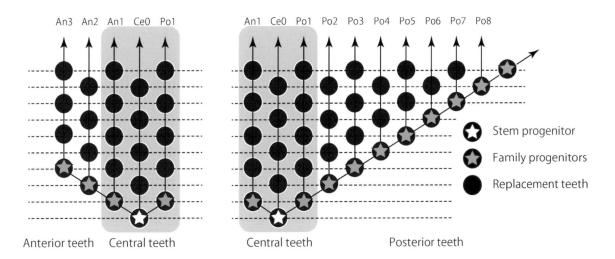

Fig. 1-8. Diagrammatic comparison of the pattern of tooth appearance in a cyprinid dentition (A-type: left) versus a cobitid dentition (right). (from Nakajima, 1987)

thus develops with alternate sweeps of even and odd replacement waves, and the teeth increase in number by one posteriorly in each even replacement wave (Fig. 1-7).

There is only one tooth in the first replacement wave, and two teeth appear at positions Po1 and An1 in the second replacement wave. The teeth that appear at these three positions, Po1, Ce0, and An1, are called the central teeth. Also there are two teeth in the third replacement wave, appearing at positions Po2 and Ce0 in cobitids, but at positions Ce0 and An2 in cyprinids. After that, the teeth increase in number posteriorly with development in cobitids. On the contrary, the teeth increase in number anteriorly in cyprinids. The anterior teeth of cyprinids that appear at positions An2 and An3 never occur in cobitids.

The anterior teeth are characteristic of cyprinids, although with an appearance pattern that varies both among and within different species. The number of tooth positions in the major row is established by the beginning of the juvenile period whereas the number of tooth positions keeps increasing during the juvenile period in cobitids. In cyprinids exhibiting a multi-row adult dentition, the tooth germs of the minor rows appear in the late post-larval period and have ankylosed to the bone by the beginning of the juvenile period (Nakajima, 1984). This never occurs in cobitids.

The pharyngeal bone of cyprinids is strongly curved, and the central teeth ankylose to the curved portion of the bone. The posterior limb of the bone extends dorsad from the curved portion to the subtemporal fossa of the skull, not directly posteriorly. That is why teeth cannot appear more posteriorly than the central teeth. Anterior teeth occur instead of posterior teeth in cyprinids, but they increase in number anteriorly with a limit, so the total number is not as high as in cobitids. The major row of cyprinid dentition commonly has just three to five teeth (Chu, 1935).

The pharyngeal bone of cobitids is elongated and arch-like and not as strongly curved as in cyprinids, and the bone grows postero-laterally during development, allowing a continuous increase in the number of teeth during the larval and juvenile periods. The shape of the pharyngeal bone seems to be reflected in both the dentition and the pattern of appearance of the tooth germs; when the pharyngeal bone is elongated and arch-like, tooth germs as a rule can appear more posteriorly than the central teeth (Fig. 1‒8). Because the catostomid pharyngeal bone is elongated and arch-like, it might be predicted that the appearance pattern of tooth germs in catostomids is the same as that in cobitids. Confirming this, Weisel (1967) and He *et al.* (1997) described a cobitid type of dentition development pattern in catostomids.

2. Morphogenesis of pharyngeal teeth in some typical cyprinid species

Since Vasnecov's (1939) descriptions of tooth development in several European cypriniform fish, a number of authors have described the development of cyprinid dentitions and the morphogenesis of their teeth. The shape of adult teeth varies depending on the species, but the first teeth are all recurved and conical. As one looks at the development of the teeth of various species, it is clear that teeth of various shapes diverge at particular stages.

Vasnecov (1939) presented a diagram of the developmental process of the pharyngeal teeth, in which the fundamental part is the "*Leuciscus* Stage" with a narrow and concave grinding surface. More recently, morphogenesis of pharyngeal teeth has been studied by SEM in the cyprinine *Cyprinus carpio* (Kodera, 1982), the hypophthalmichthyine *Hypopthalmichthys nobilis* (Nakajima and Yue, 1989), the leuciscine *Tribolodon hakonensis* (Nakajima, 1990), the labeonine *Cirrhinus molitorella* (Yue and Nakajima, 1994), the leuciscine *Mylopharyngodon piceus* (Nakajima and Yue, 1995), and the gobionine *Gnathopogon elongatus* (Sato *et al.*, 2000). Morphological changes occurring in the teeth as development progresses have been compared among these species. These changes proceed in a number of distinct stages, and cyprinine teeth pass through the greatest number of stages (Nakajima, 1998). Therefore, the morphogenesis of the pharyngeal teeth in a representative of this subfamily, a crucian carp of the genus *Carassius*, will be described in detail first herein, and the stages of morphogenesis will be established based on this. The teeth of many cyprinids turn out to correspond morphologically with particular stages of pharyngeal tooth development in crucian carp.

2-1. Morphogenesis of pharyngeal teeth in *Carassius auratus grandoculis*

Larvae and juveniles of *Carassius auratus grandoculis*, which is endemic to Lake Biwa, were reared from eggs that had been artificially fertilized in the laboratory and sampled at random. Their parent fish were captured in a ditch beside paddy fields in Otsu, Shiga Prefecture, Japan. The larval dentition is the C type, consisting of four tooth families (Nakajima, 1984). Positions A4, A3, A2, and A1 in the adult dentition correspond to positions Po1, Ce0, An1, and An2, respectively, in the larval dentition.

Developmental changes in the dentition of larvae and juveniles

Larvae of 5.6 to 7.6 mm in body length (BL) display the larval dention. At this stage, each tooth can be identified unambiguously as $_0$[Ce0] (I), $_1$[Po1] (II), $_1$[An1] (III), $_2$[Ce0] (IV), $_2$[An2] (i), and so on. In larvae that are larger than 9 mm BL, four teeth are arranged in a row, *i.e.* the adult dentition has been achieved, and the course of tooth generation at each tooth position can no longer be followed. The teeth at positions A4 (Po1), A3 (Ce0), A2 (An1), and A1 (An2) are thus referred to, respectively, as teeth A4, A3, A2, and A1 hereafter.

In a larva of 5.6 mm BL, three central teeth are ankylosed to the pharyngeal bone: $_0$[Ce0] (I), $_1$[Po1] (II), and $_1$[An1] (III). These are recurved and conical at this stage (Fig. 2-1A). In a larva of 6.0 mm BL, the three central teeth $_1$[Po1] (II), $_1$[An1] (III), and $_2$[Ce0] (IV), as well as one anterior tooth, $_2$[An2] (i), are ankylosed to the bone, and tooth $_0$[Ce0] (I) has been shed. Tooth $_2$[An2] (i) is recurved and conical, but tooth $_2$[Ce0] (IV) bears a hook at the tip and its narrow and concave grinding surface points posteriorly, with several denticles present on the margins of the grinding surface (Fig. 2-1B). In a larva of 6.8 mm BL, five central teeth, *viz.*, $_1$[Po1] (II), $_1$[An1] (III), $_2$[Ce0] (IV), $_3$[Po1] (V), and $_3$[An1] (VI), and one anterior tooth, $_2$[An2] (i), are ankylosed to the bone. Tooth $_3$[An1] (VI) is similar to tooth $_2$[Ce0] (IV). In tooth $_3$[Po1] (V), the tooth neck is slightly turned and the now outspread grinding surface points slightly laterad (Fig. 2-1C). In a larva of 6.9 mm BL, the three central teeth $_3$[Po1] (V), $_3$[An1] (VI), and $_4$[Ce0] (VII) and one anterior tooth, $_4$[An2] (iii), are ankylosed to the bone, while teeth $_0$[Ce0] (I), $_1$[Po1] (II), $_1$[An1] (III), $_2$[Ce0] (IV), and $_2$[An2] (i) have been shed. Tooth $_4$[Ce0] (VII)

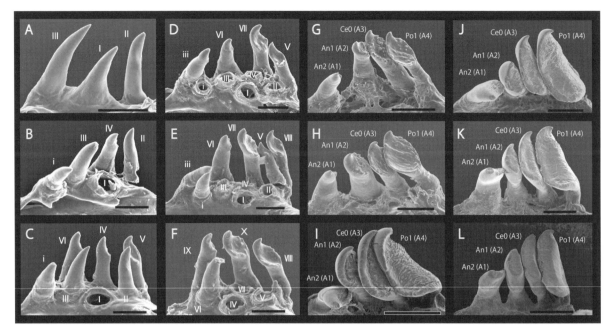

Fig. 2-1. Developmental changes of left pharyngeal dentition in larval and juvenile *Carassius auratus grandoculis*. A, lateral view of dentition at 5.6 mm BL; B, 6.0 mm BL; C 6.8 mm BL; D, 6.9 mm BL; E, 7.3 mm BL; F, 7.6 mm BL; G, 9.6 mm BL; H, 10.8 mm BL; I, 14.7 mm BL; J, 18.3 mm BL; K, 19.4 mm BL; L, 33.5 mm BL. Roman numerals: I, tooth $_0$[Ce0]; II, tooth $_1$[Po1]; III, tooth $_1$[An1]; IV, tooth $_2$[Ce0]; V, tooth $_3$[Po1]; VI, tooth $_3$[An1]; VII, tooth $_4$[Ce0]; VIII, tooth $_5$[Po1]; IX, tooth $_5$[An1]; X, tooth $_6$[Ce0]; i, tooth $_2$[An2]; ii, tooth $_3$[An3]; iii, tooth $_4$[An2]. Scales show 20 μm in A to F, 40 μm in G and H, 100 μm in I to K, and 200 μm in L.

is similar to tooth $_3$[Po1] (V) and larger than it. Tooth $_4$[An2] (iii) is similar to tooth $_3$[An1] (VI), and its narrow grinding surface points posteriorly (Fig. 2-1D). In a larva of 7.3 mm BL, the four central teeth $_3$[Po1] (V), $_3$[An1] (VI), $_4$[Ce0] (VII), and $_5$[Po1] (VIII), and also two anterior teeth, $_2$[An2] (i) and $_4$[An2] (iii), are ankylosed to the bone. In this specimen, tooth $_2$[An2] (i) remains in place without having been shed. Several denticles are present on the grinding surface, and the margins of the central teeth $_3$[Po1] (V), $_3$[An1] (VI), $_4$[Ce0] (VII), and $_5$[Po1] (VIII) are serrate. The neck of tooth $_5$[Po1] (VIII) is turned, and its grinding surface points laterad (Fig. 2-1E). In a larva of 7.6 mm BL, the four central teeth $_3$[An1] (VI), $_5$[Po1] (VIII), $_5$[An1] (IX), and $_6$[Ce0] (X) are ankylosed to the bone. The necks of teeth $_6$[Ce0] (X) and $_5$[Po1] (VIII) are both turned, and their grinding surfaces point laterad; the former lateral margin of the grinding surface has thus become the anterior margin in these teeth, and the medial margin has become the posterior one. The grinding surface of tooth $_5$[An1] (IX) points posteriorly, as is also true for teeth $_2$[Ce0] (IV) and $_3$[An1] (VI) (Fig. 2-1F).

In a larva of 9.6 mm BL, the the adult dentition is expressed, with four teeth arranged in a row. Teeth A4, A3, and A2 are central teeth and Tooth A1 is an anterior tooth. Several denticles are present on the grinding surface and margins, and the margins are serrate in all four teeth. Teeth A1 and A2 bear a hook at the tip, and the grinding surface points posteriorly. The grinding surface of tooth A1 is narrow but that of tooth A2 is broad. Teeth A3 and A4 both bear a hook at the tip and have a broad grinding surface that points laterad. The posterior margin of the grinding surface is slightly indistinct (Fig. 2-1G). In a larva of 10.8 mm BL, the grinding surfaces of teeth A1 and A2 point posteriorly, and those of teeth A3 and A4 point laterad. Several denticles are present on the grinding surface, and the margins are serrate. The grinding surface of tooth A4 has become more dilated than before (Fig. 2-1H). In teeth A2, A3, and A4 of a juvenile of 14.7 mm BL, the tooth hook is located at the medial end of the crown, and the posterior margin of the grinding surface is indistinct. The grinding surfaces of teeth A2, A3, and A4 bear many denticles, and the anterior margin is smooth but swollen. The posterior margin of these teeth is indistinct. The crown of tooth A4 is bent forward. Tooth A1 has a narrow grinding surface that points posteriorly (Fig. 2-1I). In a juvenile of 18.3 mm BL, the shapes of teeth A4, A3, and A2 are similar to those of the preceding smaller juvenile, but the teeth have become larger, so the denticles on the grinding surface are obscure (Fig. 2-1J). In a juvenile of 19.4 mm BL, the angle between the direction of the tooth axis and the grinding surface is nearly a right angle in teeth A4, A3, and A2 (Fig. 2-1K). In a juvenile of 33.5 mm BL, teeth A4, A3, and A2 are compressed antero-posteriorly, and all four teeth have become worn, with the formation of a secondary grinding surface on each of them. The dentition of this juvenile was about the same of that in adults (Fig. 2-1L).

Developmental stages of teeth in *Carassius auratus grandoculis*

The larval and juvenile teeth change morphologically from recurved cones to the adult compressed form through seven stages as follows (Figs. 2-2 and 2-3):

Stage Ca-1: Teeth $_0$[Ce0] (I), $_1$[Po1] (II), $_1$[An1] (III), and $_2$[An2] (i) present in larval dentition. All teeth recurved and conical.

Stage Ca-2: Teeth $_2$[Ce0] (IV), $_3$[An1] (VI), $_4$[An2] (iii), $_5$[An1] (IX) present in larval dentition, and tooth A2 (An1) of larvae and tooth A1 (An2) of larvae to adults present in adult dentition. Each tooth with recurved hook at tip and grinding surface situated immediately posterior to tooth hook. Grinding surface pointing posteriorly, commonly narrow, but broad in tooth A2 of post-larva.

Stage Ca-3: Teeth $_3$[Po1] (V) and $_4$[Ce0] (VII) present in larval dention. Tooth necks slightly turned, grinding surface pointing slightly laterally with denticles.

Stage Ca-4: Teeth $_5$[Po1] (VIII) and $_6$[Ce0] (X) present in larval dentition, and A4 (Po1) and A3 (Ce0) of larvae, present in adult dentition. Grinding surface pointing laterad, its lateral margin thus becoming anterior, and medial margin posterior, with anterior margin curving forward. Grinding surface broader than in preceding stage, bearing many denticles.

Stage Ca-5: Teeth A4 (Po1), A3 (Ce0), and A2 (An1) of juveniles present in adult dentition, with necks more twisted than before and hook located at medial end of crown. Anterior margin of grinding surface swollen, and posterior margin indistinct. Grinding surface of some teeth such as A4 and A3 dilated, sputlate.

Stage Ca-6: Teeth A4 (Po1), A3 (Ce0), and A2 (An1) of juveniles and adults present with tooth hook located at medial end of crown. Angle between direction of tooth axis and grinding surface nearly a right angle. Teeth compressed antero-posteriorly. Tooth crown of all teeth worn with formation of secondary grinding surface.

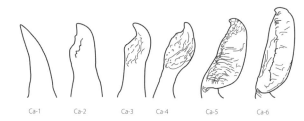

Fig. 2-2. Developmental stages of right pharyngeal teeth in *Carassius auratus grandoculis*. Refer to Table 2-1 for full details of stages Ca-1 to Ca-6.

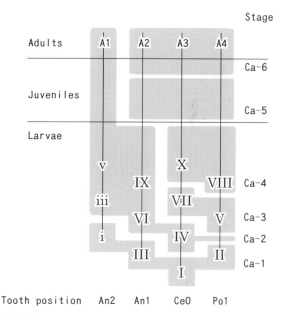

Fig. 2-3. Developmental stages of each tooth in the larval and adult left dentitions of *Carassius auratus grandoculis*. Refer to Fig. 2-1 for explanation of Roman numerals.

2-2. Morphogenesis of pharyngeal teeth in *Acrossocheilus parallens*

Larvae and juveniles of *Acrossocheilus parallens* reared from eggs artificially fertilized in the laboratory were sampled at random. Their parent fish were collected in Guangdong, China, and bred in an aquarium. The tooth number of the A row in the adult dentition is five, so Positions A5, A4, A3, A2, and A1 in the adult dentition correspond, respectively, to positions Po1, Ce0, An1, An2, and An3 in the larval dentition.

Developmental changes in the larval dentition or A-row teeth

Larvae of 8 mm to 15 mm BL display the larval dentition, each tooth being identified as $_0$[Ce0] (I), $_1$[Po1] (II), $_1$[An1] (III), $_2$[Ce0] (IV), $_2$[An2] (i), and so on. In

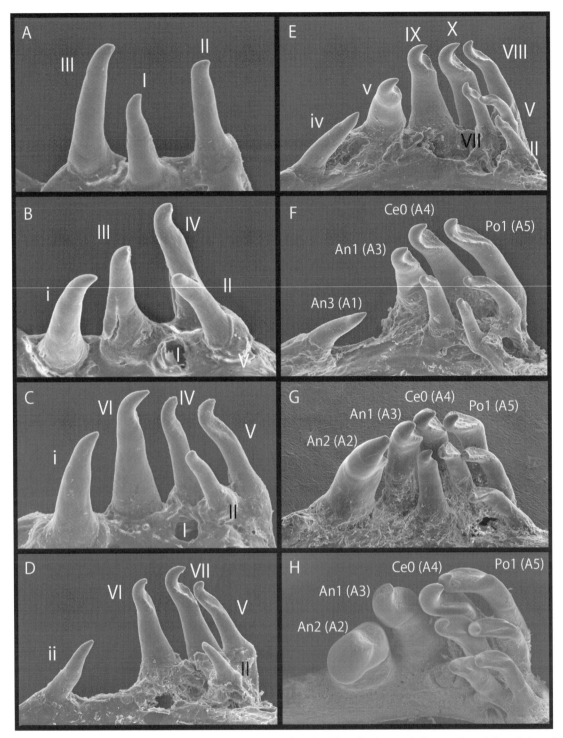

Fig, 2-4. Developmental changes of left pharyngeal dentition in larval and juvenile *Acrossocheilus parallens*. A, lateral view of dentition at 8.1 mm BL; B, 8.9 mm BL; C, 10.9 mm BL; D, 12.0 mm BL; E, 15.6 mm BL; F, 19.1 mm BL; G, 31.1 mm BL; H, 103.0 mm BL. Refer to Fig. 2-1 for explanation of Roman numerals.

larvae of more than 15 mm BL, a transitional dentition is expressed and the arrangement of the lateral teeth is irregular, thus making it difficult to identify and trace the fate of each tooth (Fig. 2-4).

Teeth at position Po1 (A5): Tooth $_1$[Po1] (II) is recurved and conical. Teeth $_3$[Po1] (V) and $_5$[Po1] (VIII) bear a large, recurved hook at the tip and a posteriorly pointing concave grinding surface at the base of the hook, as well as some denticles on the grinding surface and its margins. Tooth A5 of a larva of 19.1 mm BL has a broad, concave grinding surface with sharp, jutting medial and lateral margins. In tooth A5 of juveniles of more than 31.1 mm BL and adults, the tooth crown is bent forward, and the grinding surface points dorsad, wider than before. There is a ridge on the grinding surface. In adults, both the lateral and medial margins are swollen.

Teeth at position Ce0 (A4): Teeth $_0$[Ce0] (I) and $_2$[Ce0] (IV) are recurved and conical. Tooth $_4$[Ce0] (VII) and $_6$[Ce0] (X) have a broad, concave grinding surface with sharp, jutting medial and lateral margins. They bear some denticles on the grinding surface and its margins. Tooth A4 of a juvenile of 19.1 mm BL is larger than before and similar to tooth $_6$[Ce0] (X), but the denticles on the grinding surface are obscure. There is a ridge on the grinding surface. In tooth A4 of juveniles of more than 31.1 mm BL and adults, the concave grinding surface is broadened and points dorsad. In adults, both the lateral and medial margins are swollen. In tooth A4 of adults, the tooth crown is bent forward, and the grinding surface points dorsad.

Teeth at position An1 (A3): Tooth $_1$[An1] (III) is recurved and conical. Tooth $_3$[An1] (VI) has a small grinding surface. Tooth $_5$[An1] (IX) has a broad, concave grinding surface at the base of the tooth hook; both margins jut out and are sharp. Tooth A3 of juveniles of more than 19.1 mm BL has a broad, concave grinding surface with sharp, jutting medial and lateral margins. In adults, the tooth crown becomes large, and both the lateral and medial margins and the hook become swollen.

Teeth at position An2 (A2): Teeth $_2$[An2] (ii), $_4$[An2] (iv), and A2 of larvae are recurved and conical. Tooth A2 of juveniles of more than 15.6 mm BL and adults has a concave grinding surface pointing posteriorly. In adults, the tooth crown becomes large and both the lateral and medial margins swell.

Teeth at position An3 (A1): Tooth A1 is conical in larvae, juveniles, and adults.

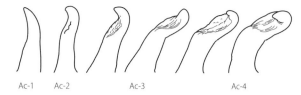

Fig. 2-5. Developmental stages of right pharyngeal teeth in *Acrossocheilus parallens*. Refer to Table 2-1 for full details of stages Ac-1 to Ac-4.

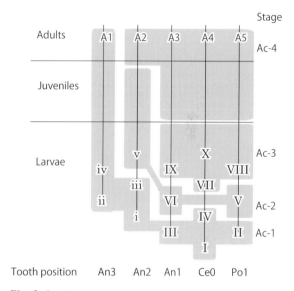

Fig. 2-6. Developmental stages of each tooth in left larval dentition and A row of left adult dentition of *Acrossocheilus parallens*. Refer to Fig. 2-1 for explanation of Roman numerals.

Developmental stages of teeth in *Acrossocheilus parallens*

The larval and juvenile teeth change morphologically from recurved and conical to the adult form through four stages as follows (Figs. 2-5 and 2-6):

Stage Ac-1: Teeth $_0$[Ce0] (I), $_1$[Po1] (II), $_1$[An1] (III), $_2$[Ce0] (IV), $_2$[An2] (i), $_3$[An3] (ii), $_4$[An2] (iii), and $_5$[An3] (iv) in lrvae, and A1 (An3) in larvae, juveniles, and adults. Recurved and conical.

Stage Ac-2: Teeth $_3$[Po1] (V), $_3$[An1] (VI), and $_6$[An2] (v) of larvae, and tooth A2 (An2) in larvae and juveniles, with hook at the tip and a narrow, concave grinding surface immediately posterior to the hook; some with denticles on the grinding surface.

Stage Ac-3: Teeth $_4$[Ce0] (VII), 5[Po1] (VIII), 5[An1] (IX), $_6$[Ce0] (X), as well as teeth A5 (Po1), A4 (Ce0), and A3 (An1) in larvae. Grinding surface expanded, now concave and broad with sharp, jutting medial and lateral margins, grinding surface smooth or with denticles.

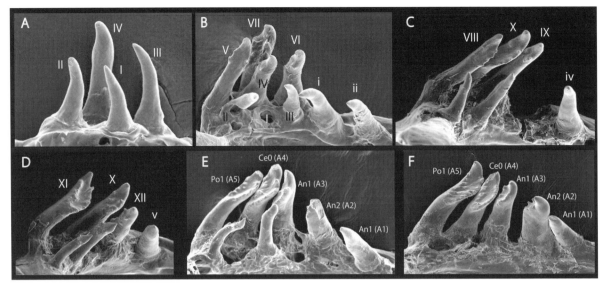

Fig. 2-7. Developmental changes of right pharyngeal dentitions in larval and juvenile *Gnathopogon elongatus*. A, lateral view of dentition at 6.7 mm BL; B, 8.8 mm BL; C, 11.6 mm BL; D, 11.9 mm BL; E, 13.0 mm BL; F, 17.8 mm BL. Refer to Fig. 2-1 for explanation of Roman numerals.

Stage Ac-4: Teeth A5 (Po1), A4 (Ce0), and A3 (An1) in juveniles and adults and tooth A2 (An2) in adults. Tooth crown enlarged, medial and lateral margins swollen into ridges, bulge or ridge present on grinding surface, and in teeth A5 (Po1) and A4 (Ce0), tooth crown bent forward, grinding surface pointing dorsad.

2-3. Morphogenesis of pharyngeal teeth in *Gnathopogon elongatus*

Tooth morphogensis in *G. elongatus* was reported in detail by Sato *et al.* (2000). The developmental process is divided into six stages based on their descriptions (Figs. 2-7, 2-8, and 2-9).

Stage Gn-1: Teeth $_0$[Ce0] (I), $_1$[Po1] (II), $_1$[An1] (III), $_2$[An2] (i), $_3$[An3] (ii), $_4$[An2] (iii), and $_5$[An3] (iv), as well as tooth A1 of a juvenile of 11.9 mm BL. Recurved and conical.

Stage Gn-2: Teeth $_2$[Ce0] (IV), $_3$[An1] (VI), and $_6$[An2] (v) in larvae, tooth A2 (An2) in juveniles and adults, and tooth A1 in juveniles and adults of more than 17.8 mm BL, with hook at the tip and a concave, posteriorly pointing grinding surface. Tooth $_2$[Ce0] (IV) remaining recurved and conical in some individuals.

Stage Gn-3: Teeth $_3$[Po1] (V), $_4$[Ce0] (VII), $_5$[An1] (IX), and $_7$[An1] (XII) in larvae, as well as tooth A3 (An1) in juveniles, and adults. Tooth neck slightly turned and concave grinding surface pointing slightly laterad; lateral (anterior) and medial (posterior) margins formed alike.

Stage Gn-4: Teeth $_5$[Po1] (VIII), $_6$[Ce0] (X), and $_7$[Po1]

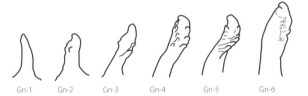

Fig. 2-8. Developmental changes of right pharyngeal teeth in *Gnathopogon elongatus*. Refer to Table 2-1 for full details of stages Gn-1 to Gn-6.

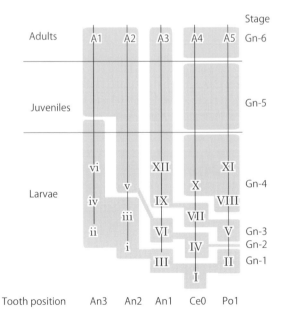

Fig. 2-9. Developmental stages of each tooth in left larval dentition and A row of left adult dentition of *Gnathopogon elongatus*. Refer to Fig. 2-1 for explanation of Roman numerals.

(XI), as well as teeth A5 and A4 of larvae. Tooth neck turned, grinding surface pointing laterad, anterior margin of grinding surface curving forward, posterior margin becoming slightly indistinct. Line of denticles on both anterior and posterior margins of concave grinding surface; in tooth A5, crown bent forward.

Stage Gn-5: Teeth A5 (Po1) and A4 (Ce0) in juveniles. Neck more turned than before. Hook located at medial end of crown. Anterior margin of grinding surface clearly expressed and curved forward, posterior margin more indistinct than before. Grinding surface spatulate with denticles. In tooth A5 (Po1), crown bent forward.

Stage Gn-6: Teeth A5 (Po1) and A4 (Ce0) in adults. Tooth crown compressed in antero-posterior direction, and grinding surface relatively narrow and sparsely denticulate.

2-4. Common stages of tooth development

As was summarized above, the developmental stages of the teeth at each position have been observed in three species of cyprinids, *Carassius auratus grandoculis, Acrossocheilus parallens,* and *Gnathopogon elongatus.* The initial teeth in all three species are recurved and conical, and the following teeth have a narrow and concave grinding surface. The first two stages (Stages Ca-1 and Ca-2, Stages Ac-1 and Ac-2, and Stages Gn-1 and Gn-2, respectively) are same in all three fish's teeth, so these stages will be called common stages 1 and 2, respectively, in the rest of this report.

In *Acrossocheilus*, while the grinding surface points posteriorly, the shape of the tooth crown changes as the teeth are replaced. At Stage Ac-3, the grinding surface enlarges, and at Stage Ac-4, the tooth crown becomes larger and gains a longitudinal ridge. In teeth A5 and A4, the margins swell into ridges, and in teeth A3 and A2, both the hook and margins swell and the tooth crown becomes massive.

In *Carassius* and *Gnathopogon*, the morphological changes the teeth undergo are different from those in *Acrossocheilus*. In the former two genera, the tooth neck twists and the grinding surface eventually points laterad. The original lateral margin becomes the anterior one and is curved forward and the grinding surface becomes enlarged (Stages Ca-4 and Gn-4). This stage is common Stage 3 for these two genera. At the next step, the posterior margin (original medial margin) becomes indistinct (Stages Ca-5 and Gn-5). In *Gnathopogon*,

the tooth crown is compressed in an antero-posterior direction, and the grinding surface becomes narrow (Stage Gn-6). These stages belong to common Stage 4 for these two genra. No additional common stage can be recognized in *Gnathopogon*.

At the next stage in *Carassius*, the tooth axis comes to make nearly a right angle with the grinding surface (Stage Ca-6), but in *Gnathopogon*, these axes still cross aslant, as in teeth at common Stage 4 (Stage Gn-6) (Table 2-1).

2-5. Developmental changes of the central teeth in some cyprinids

Although the shape of the central teeth at positions Po1, Ce0, and An1, which are the most posterior three positions in the A row, undergoes changes, in the anterior teeth at position An2 and An3, Stages 1 and 2 of tooth development are maintained through the juvenile and even adult stages. Therefore, in this section only developmental changes of the central teeth are compared among various cyprinids.

2-5-1. *Opsariichthys bidens* (Fig. 2-10)

The central teeth change morphologically being from recurved and conical (Stage 1: $_0$[Ce0] (I), $_1$[Po1] (II), and $_1$[An1] (III)) to having a narrow and concave grinding surface that points posteriorly (Stage 2: $_2$[Ce0] (IV)) or slightly laterad (Stage 2: $_3$[Po1] (V), $_3$[An1] (VI), $_4$[Ce0] (VII), $_5$[Po1] (VIII), $_5$[An1] (IX), and $_6$[Ce0] (X)). In the adult dentition, the central teeth have a narrow and concave grinding surface that points slightly laterad. In the most posterior two teeth of adults, Po1 and Ce0, the tooth crown is bent forward.

2-5-2. *Spinibarbus sinensis* (Fig. 2-11)

The central teeth change from being recurved and conical (Stage 1: $_1$[Po1] (II) and $_2$[Ce0] (IV)) into adult teeth much as those of *Acrossocheilus parallens* do, through two stages (Stage Ac-2: $_4$[Ce0] (VII), $_5$[An1] (IX), and the other teeth at position An1 from larvae to adults; Stage Ac-3: the other teeth at positions Po1 and Ce0 in juveniles and adults). In adult fish, the center of the grinding surface rises and forms a bulge. A continuous ridge with both the medial and lateral margins of the grinding surface as well as the tooth hook swollen forms a marginal ridge. A shallow, U-shaped groove forms between the bulge and the marginal ridge. In the most posterior two teeth of adults, A5 (Po1) and A4 (Ce0),

Table 2-1. Corresponding developmental stages of pharyngeal teeth in three cyprinid genera, *Carassius*, *Gnathopogon*, and *Acrossocheilus*.

Common stages	Stages in *Carassius auratus grandoculis*	Stages in *Gnathopogon elongatus elongatus*	Stages in *Acrossocheilus parallens*
Stage 1 Recurved and conical	Stage Ca-1 Recurved and conical	Stage Gn-1 Recurved and conical	Stage Ac-1 Recurved and conical
Stage 2 Grinding surface pointing backward	Stage Ca-2 Narrow and concave grinding surface pointing backward	Stage Gn-2 Narrow and concave grinding surface pointing backward	Stage Ac-2 Narrow and concave grinding surface pointing backward
	Stage Ca-3 Broad and concave grinding surface pointing slightly laterad	Stage Gn-3 Broad and concave grinding surface pointing slightly laterad	Stage Ac-3 Broad and concave grinding surface pointing backward
			Stage Ac-4 Various structures present on grinding surface and its margins
Stage 3 Grinding surface pointing laterad	Stage Ca-4 Broad and concave grinding surface pointing laterad	Stage Gn-4 Broad and concave grinding surface pointing laterad	
Stage 4 Posterior margin indistinct	Stage Ca-5 Broad and concave grinding surface pointing laterad Posterior margin indistinct	Stage Gn-5 Broad and concave grinding surface pointing laterad Posterior margin indistinct	
		Stage Gn-6 Tooth crown slightly compresed	
Stage 5 Direction of tooth axis makes right angle with grinding surface	Stage Ca-6 Direction of tooth axis makes right angle with grinding surface		

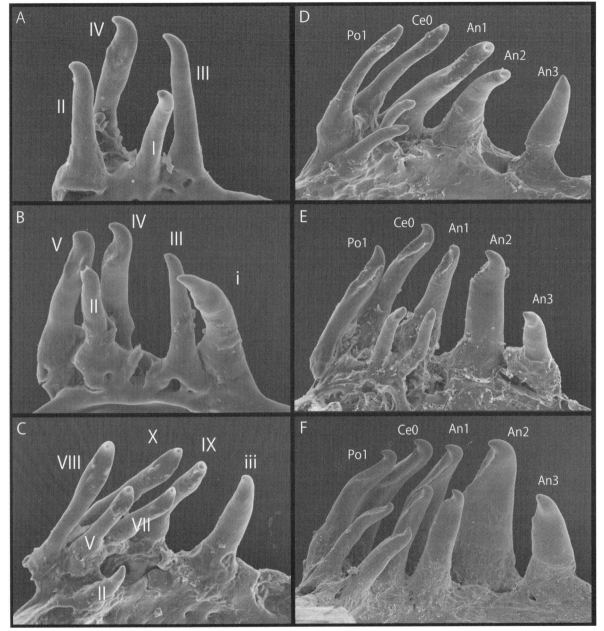

Fig. 2-10. Developmental changes of right central teeth in *Opsariichthys bidens*. A, lateral view of dentition at 6.2 mm BL; B, 7.4 mm BL; C, 9.2 mm BL; D, 12.5 mm BL; E, 13.9 mm BL; F, 46.0 mm BL. Refer to Fig. 2-1 for explanation of Roman numerals.

the tooth crown is bent forward.

2-5-3. *Megalobrama amblycephala* (Fig. 2-12)

The central teeth are initially recurved and conical (Stage 1), but these are replaced by teeth with a narrow and concave grinding surface that points posteriorly (Stage 2). The tooth neck turns, and the grinding surface comes to be directed laterad as in *Carassius* and *Gnathopogo*n (Stage 3). The teeth then become compressed, and a hamulus groove develops on the posterior side of the crown at the base of the tooth hook.

2-5-4. *Hypophthalmichthys nobilis* (Fig. 2-13)

The developmental changes of the pharyngeal teeth were reported by Nakajima and Yue (1989). The initial teeth are recurved and conical (Stage 1: $_0$[Ce0] (I), $_1$[Po1] (II), and $_1$[An1] (III)), and in the next stage the teeth have a concave and narrow grinding surface that points

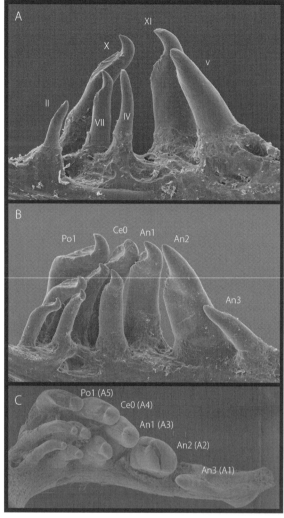

Fig. 2-11. Developmental changes of right central teeth in *Spinibarbus sinensis*. A, lateral view at 16.9 mm BL; B, 21.6 mm BL; C, dorsal view at 164.0 mm BL. Refer to Fig. 2-1 for explanation of Roman numerals.

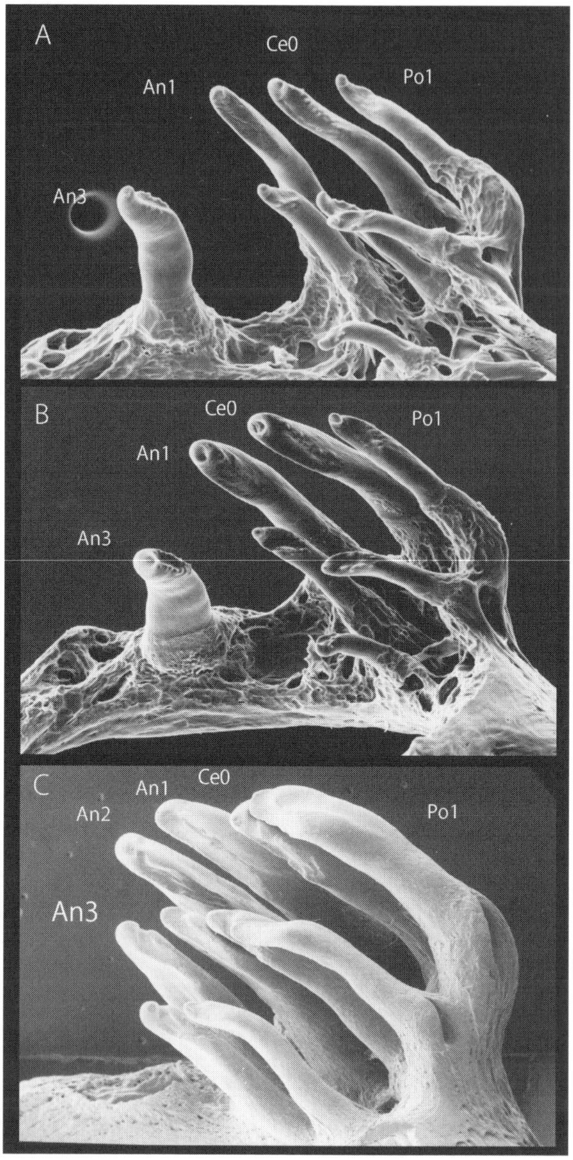

Fig. 2-12. Developmental changes of left central teeth in *Megalobrama amblycephala*. A, lateral view of dentition at 19.1 mm BL; B, 19.7 mm BL; C, 225.0 mm BL.

posteriorly (Stage 2: $_2$[Ce0] (IV), $_3$[Po1] (V), and $_3$[An1] (VI)). From the following stages the tooth neck turns and the grinding surface points laterad, but the posterior margin (original medial margin) does not become indistinct. A hook is seen at the medial end of the crown in juveniles, but it disappears in adults.

2-5-5. *Rhodeus ocellatus* and *Acheilognathus rhombeus* (Fig. 2-14)

In both species the adult dentition is achieved already in larvae. The grinding surface points laterad and both the anterior and posterior margins are distinct (Stage 3). The teeth are compressed and the grinding surface is narrow. Some denticles are present on the margins, and the margins are serrate in larvae. In adult *A. rhombeus*, the anterior magin is serrate and the posterior margin is smooth. A hamulus groove is formed on the posterior side of the crown at the base of the tooth hook. In adults of *Rhodeus ocellatus*, both the anterior and posterior margins are smooth.

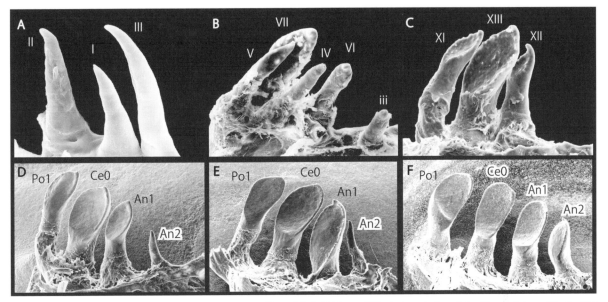

Fig. 2-13. Developmental changes of right central teeth in *Hypophthalmichthys nobilis*. A, lateral view of dentition at 8.4 mm BL; B, 10.5 mm BL; C, 12.9 mm BL; D, 18.2 mm BL; E, 23.0 mm BL; F, 34.0 BL. Refer to Fig. 2-1 for explanation of Roman numerals.

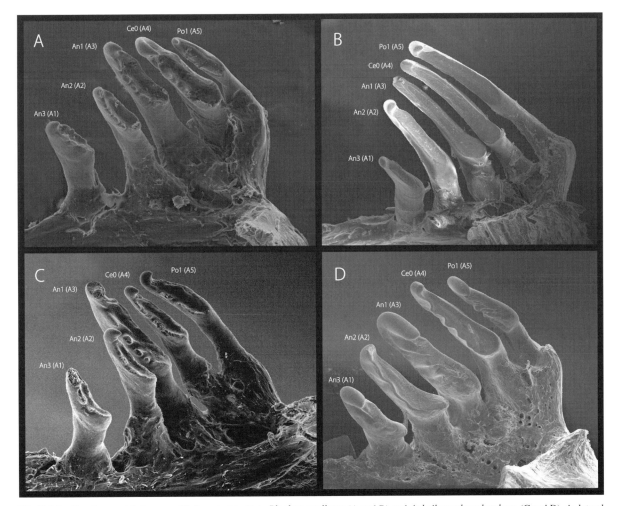

Fig. 2-14. Developmental changes of left central teeth in *Rhodeus ocellatus* (A and B) and *Acheilognathus rhombeus* (C and D). A, lateral view of dentition in a larva at 9.9 mm BL; B, adult at 41.2 mm BL; C, larva at 13.0 mm BL; D, adult at 70.1 mm BL.

2. Morphogenesis of pharyngeal teeth in some typical cyprinid species

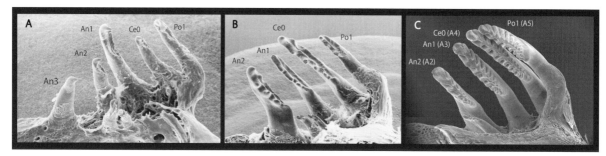

Fig. 2-15. Developmental changes of left central teeth in *Ctenopharyngodon idella*. A, lateral view of dentition at 18.7 mm BL; B, 28.2 mm BL; C, 107.0 mm in BL.

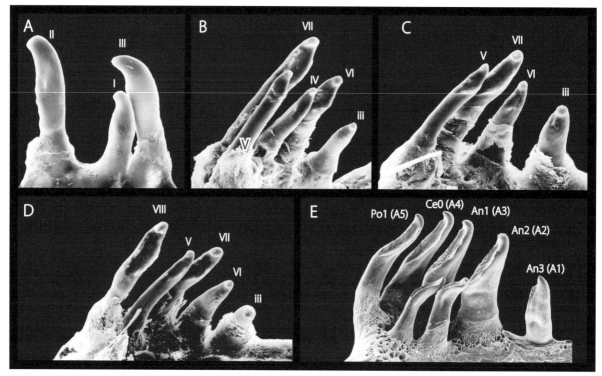

Fig. 2-16. Developmental changes of right central teeth in *Tribolodon hakonensis*. A, lateral view of dentition at 11.5 mm BL; B, 15.2 mm BL; C, 15.2 mm BL; D, 15.3 mm BL; E, 120.0 mm BL. Refer to Fig. 2-1 for explanation of Roman numerals.

2-5-6. *Ctenopharyngodon idella* (Fig. 2-15)

The initial teeth are recurved and conical (Stage 1). In later teeth of larvae to adults, the grinding surface points laterad and both margins are distinct (Stage 3). Some denticles are present on both the anterior and posterior margins and both margins are serrate in larvae and juveniles. Both the anterior and posterior sides of the tooth crown are obliquely grooved in teeth of adults.

2-5-7. *Tribolodon hakonensis* (Fig. 2-16)

The developmental changes of the teeth were reported by Nakajima (1990). The central teeth are initially at Stage 1 ($_0$[Ce0] (I), $_1$[Po1] (II), and $_1$[An1] (III)) or Stage 2 ($_2$[Ce0] (IV) and $_3$[An1] (VI)). In $_4$[Ce0] (VII) and $_5$[Po1] (VIII), the grinding surface points slightly laterad (Stage 2). In all the central teeth of juveniles and adults, the grinding surface points laterad (Stage 3). In adults, the posterior margins are distinct (Stage 3) or indistinct (Stage 4).

2-5-8. *Cirrhinus molitorella* (Fig. 2-17)

The developmental changes of the teeth were reported by Yue and Nakajima (1994). The central teeth start as Stages 1 and 2, but then the tooth neck turns, and the grinding surface points laterad (Stage 3). The grinding surface then become dilated and spatulate, and the posterior margin (original medial margin) becomes indistinct (Stage 4). The teeth become compressed and intensely worn, and a secondary grinding surface is formed.

2-5-9. *Pseudorasbora parva* (Fig. 2-18)

The initial teeth are recurved and conical (Stage 1: $_0$[Ce0] (I), $_1$[Po1] (II), and $_1$[An1] (III)). Tooth $_2$[Ce0] (IV) has a concave grinding surface that points backward (Stage 2). In juveniles, the grinding surface points laterad and the posterior margin become indistinct (Stage 4). In adults, the central teeth are compressed, have a narrow and concave grinding surface, and arise from dentigerous surface of triangular prominence.

2-5-10. *Sarcocheilichthys sinensis* (Fig. 2-19)

The initial teeth are recurved and conical (Stage 1: $_0$[Ce0] (I), $_1$[Po1] (II), and $_1$[An1] (III), and $_2$[Ce0] (IV)) and the following teeth have a narrow grinding surface that points posteriorly (Stage 2: $_3$[Po1] (V), $_3$[An1] (VI), and $_4$[Ce0] (VII)). In $_5$[Po1] (VIII) and $_5$[An1] (IX), the tooth neck twists slightly (Stage 2). In juveniles the neck turns more, and the grinding surface points laterad (Stage 3); in addition, the anterior margin (original lateral margin) curves forward and becomes strong while the posterior margin becomes indistinct (Stage 4). In the anterior two central teeth (Ce0 and An1), the anterior margin swells and a groove is formed along the anterior margin.

2-5-11. *Tinca tinca* (Fig. 2-20)

The initial teeth are recurved and conical (Stage 1: $_0$[Ce0] (I), $_1$[Po1] (II), $_1$[An1] (III), and $_2$[Ce0] (IV)). Teeth $_3$[Po1] (V), $_3$[An1] (VI), and $_4$[Ce0] (VII) have a concave and narrow grinding surface that points backward, and in $_5$[Po1] (VIII) and $_5$[An1] (IX) the grinding surface points slightly laterad (Stage 2). In larvae, the grinding surface points laterad (Stage 3). In juveniles, the grinding surface points laterad and posterior margin become indistinct (Stage 4). Although the most posterior two central teeth (Po1 and Ce0) of adults stay at Stage 4, in the most anterior central tooth (An1) of adults, the direction of the tooth axis makes a right angle with the grinding surface (Stage 5).

2-5-12. *Mylopharyngodon piceus* (Fig. 2-21)

The developmental changes of the teeth were reported by Nakajima and Yue (1995). The central teeth are initially of Stages 1 and 2, but thereafter the tooth neck twists and the grinding surface faces laterad (Stage 3). Then the posterior margin (original medial margin) becomes indistinct, and the grinding surface expands and becomes spatulate (Stage 4). Next the tooth axis makes a nearly right angle with the grinding surface (Stage 5), the tooth crown swells, and the teeth become depressed. A shallow groove runs along the anterior margin, and a trace of the weak tooth hook remain at the medial end of the crown. In the adult, the shallow groove disappears, and teeth become massive and round.

2-5-13. *Cyprinus carpio* (Fig. 2-22)

Although common carp, *Cyprinus carpio*, have only three teeth in the major row, their larval dentition is of the D type with four tooth families. The number of tooth families is not necessarily consistent with the number of major-row teeth. Common carp keep four major-row teeth up into the early juvenile period. Subsequently, one tooth, which is ankylosed at position Po1, is shed without being replaced, and three teeth remain in the major row thereafter (Kodera, 1982). That is why the adult position A3 is actually position Ce0. Teeth A3 and A2 are central teeth and Tooth A1 is an anterior tooth. The initial central teeth are recurved and conical (Stage 1), but the following teeth bear a narrow, concave grinding surface that points posteriorly (Stage 2). The tooth neck of the central teeth then twists, the lateral margin becomes the anterior one, and the medial margin becomes the posterior one (Stage 3). The posterior margin becomes indistinct, the grinding surface expands, and tooth crown become spatulate, with a concave, grooved grinding surface (Stage 4). The tooth axis intersects the grinding surface almost squarely (Stage 5). Then tooth A2 (An1), becomes massive, and the dorsal side of the tooth crown becomes flat. A trace of the tooth hook remains at the medial end of the crown. A groove runs from the base of the tooth hook along the anterior margin on the dorsal side of the crown, and the grooves increase in number with growth of the fish. Dorsal views of tooth A2 were drown by Kodera (1982), and the change of direction of the grinding surface and is shown here (Fig. 2-23).

2-6. Developmental stages in cyprinid teeth

Morphological changes of the central teeth have been newly observed in many cyprinids and reviewed based on previous descriptions in some other cyprinids.

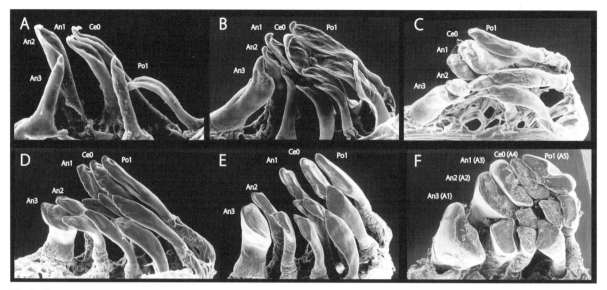

Fig. 2-17. Developmental changes of left central teeth in *Cirrhinus molitorella*. A, lateral view of dentition at 8.8 mm BL; B, larva of 10.0 mm BL; C, larva of 15.0 mm BL; D, larva of 19.7 mm BL; E, juvenile of 27.4 mm BL; F, adult of 170.0 mm BL.

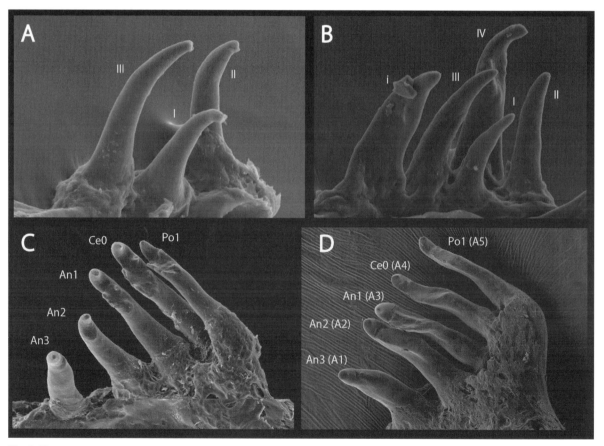

Fig. 2-18. Developmental changes of left central teeth in *Psedorasbora parva*. A, lateral view of dentition at 5.7 mm BL; B, 6.6 mm BL; C, 13.2 mm BL; D, 65.7 mm BL. Refer to Fig. 2-1 for explanation of Roman numerals.

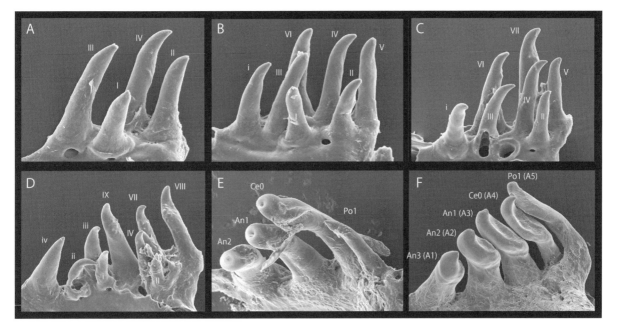

Fig. 2-19. Developmental changes of left central teeth in *Sarcocheilichthys sinensis*. A, lateral view of dentition at 8.1 mm BL; B, 9.5 mm BL; C, 10.6 mm BL; D, 12.2 mm BL; E, 19.1 mm BL; F, 70.0 mm BL. Refer to Fig. 2-1 for explanation of Roman numerals.

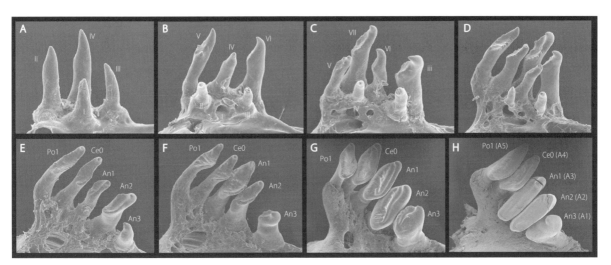

Fig. 2-20. Developmental changes of right central teeth in *Tinca tinca*. A, lateral view of dentition at 7.8 mm BL; B, 8.8 mm BL; C, 9.8 mm BL; D, 11.4 mm BL; E, 14.3 mm BL; F, 21.3 mm BL; G, 126.0 mm BL; H, 154.0 mm BL. Refer to Fig. 2-1 for explanation of Roman numerals.

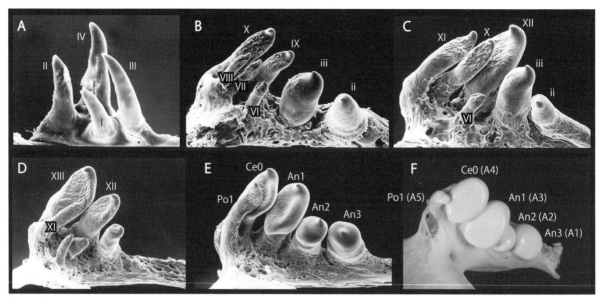

Fig. 2-21. Developmental changes of right central teeth in *Mylopharyngodon piceus*. A, latral view of dentition at 8.8 mm BL; B, 16.9 mm BL; C, 21.8 mm BL; D, 30.0 mm BL; E, 37.9 mm BL; F, 144.3 mm BL. Refer to Fig. 2-1 for explanation of Roman numerals.

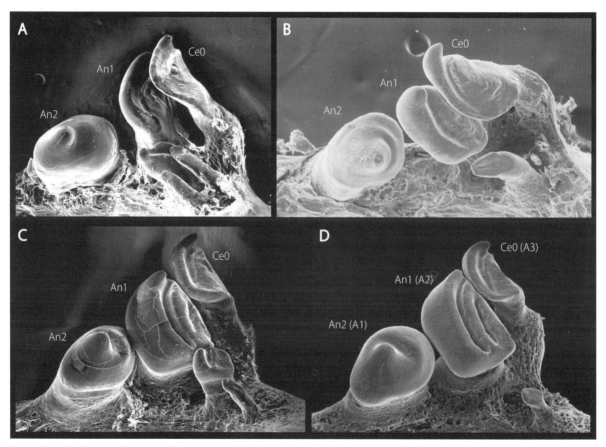

Fig. 2-22. Developmental changes of left central teeth in *Cyprinus carpio*. A, lateral view of dentition at 27.6 mm BL; B, latero-dorsal view of dentition at 42.1 mm BL; C, 84.8 mm BL; D, 109.3 mm BL.

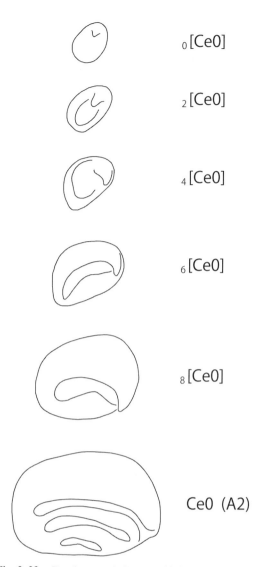

Fig. 2-23. Developmental changes of left teeth at position An1 (A2) in *Cyprinus carpio*. (After Kodera, 1982)

Table 2-2. The final developmental stage attained by the central teeth in various species of Cyprinidae.

Taxon	Final stage
Acrossocheilus parallens	Stage 2
Spinibarbus sinensis	Stage 2
Opsariichthys bidens	Stage 2
Megalobrama amblycephala	Stage 3
Hypophthalmichthys nobilis	Stage 3
Acheilognathus rhombeus	Stage 3
Rhodeus ocellatus	Stage 3
Ctenopharyngodon idella	Stage 3
Tribolodon hakonensis	Stage 3/4
Cirrhinus molitorella	Stage 4
Pseudorasbora parva	Stage 4
Gnathopogon elongatus	Stage 4
Sarcocheilichthys sinensis	Stage 4
Tinca tinca	Stage 4/5
Mylopharyngodon piceus	Stage 5
Carassius auratus grandoculis	Stage 5
Cyprinus carpio	Stage 5

The initial teeth are recurved and conical and the subsequent teeth have a narrow and concave grinding surface commonly bearing several denticles (Stage 1). The grinding surface points backward or slightly laterad in the central teeth of *Acrossocheilus parallens*, *Spinibarbus sinensis*, and *Opsariichthys bidens*, and the anterior teeth of almost all species (Stage 2).

The tooth neck twists and the grinding surface comes to point laterad in the other cyprinid central teeth. In *Megalobrama amblycephala*, *Hypophthalmichthys nobilis*, *Acheiloganathus rhombeus*, *Rhodeus ocellatus*, and *Ctenopharyngodon idella*, the posterior margin (original medial margin) does not become indistinct (Stage 3), but in *Cirrhinus molitorella*, *Gnathopogon elongatus*, *Pseudorasbora parva*, *Sarcocheilichthys sinensis*, *Tinca tinca*, *Mylopharyngodon piceus*, *Cyprinus carpio*, and *Carassius auratus grandoculis*, it does become indistinct (Stage 4). In all the central teeth, or a part of them in latter four species, the tooth axis intersects the grinding surface almost squarely (Stage 5) (Table 2-2).

3. Descriptions of the pharyngeal dentition of cyprinid subfamilies

In this section, the pharyngeal dentition, the morphology of each tooth, and the tooth number of each row as expressed by a dental formula, is given for species from every subfamily of Cyprinidae. The classification of the cyprinid subfamilies follows Chen *et al.* (1998).

Cyprinid pharyngeal teeth are classified into various types based on developmental stages, but since variation exists, they are not classifiable by this criterion alone. In addition to developmental stage, categorization was carried out based on the structure of the grinding surface, the state of its margins, the presence of grooves on the grinding surface, and so on.

Dental formula

The dental formula is based on the number of teeth of each row on either side of the throat. The left side of the formula expresses the left dentition, and the right side expresses the right dentition. If the dental formula is 2.4.5‒4.3.2, for example, the number of teeth in the right A row is four, that in the right B row is three, and that in the right C row is two, whereas the number of teeth in the left A row is five, that in the left B row is four, and that in the left C row is two. When there is variation in the number of teeth, each term is expressed as, for example, "(2,3)" or "2(3)". In the former, "(2,3)", both numbers of teeth occur at almost the same frequency; in the latter, the number in parentheses is exhibited by a minority of specimens.

Morphorogical types based on developmental stages of teeth

Pharyngeal teeth change form with growth through a number of stages. The precise morphological changes differ among taxa, and so does the timing of thses changes. Nonetheless, teeth can be typified by developmental stage (Table 3‒1).

Type-1 teeth display the characteristics of developmental Stage 1 and are recurved and conical. Although there is a hook, there is no grinding surface.

Type-2 teeth display the characteristics of developmental Stage 2, in which the grinding surface points backward, and the tooth hook is located at the anterior end of the tooth crown. Type 2 can be divided into several subtypes. In Type 2‒0 teeth, the grinding surface is not concave. Type 2‒1 teeth have a concave and narrow grinding surface. In Type 2‒2 teeth, the grinding surface is concave and outspread. There are structures, for example a U-shaped groove or a bulge, etc. on the grinding surface in some fish. The grinding surface still points backward in Type 2‒0, Type 2‒1, and Type 2‒2 teeth, but in Type 2‒3, the grinding surface points slightly laterad.

Type-3 teeth display the charcteristics of developmental Stage 3, in which the grinding surface points laterad, both the anterior and posterior margins of the grinding surface are distinct with sharp edges, and the tooth hook is located at the medial end of the crown. Type-3 can be divided into two subtypes. Type 3‒0 has a concave and narrow grinding surface that points laterad. Type 3‒1 has a broad grinding surface that also points laterad.

Type-4 teeth display the charcteristics of developmental Stage 4, in which the grinding surface points laterad and the posterior margin is indistinct. In Type 4‒0 teeth, the posterior margin is slightly indistinct, and the grinding surface is usually concave and narrow. In Type 4‒1 teeth, the posterior margin is significant indistinct, or the grinding surface and the posterior side of the crown are united. The grinding surface is dilated like a spatula, or the tooth crown is compressed antero-posteriorly in some fish.

Type-5 teeth display the characteristics of developmental Stage 5, in which the tooth axis makes a right angle with the grinding surface. In Type 5‒0 teeth, a shallow groove runs close to the anterior margin on the grinding surface. This groove becomes deep and increases in number in Type 5‒1 teeth. Teeth of Types 5‒0 and 5‒1 are usually massive. In Type 5‒2 teeth, the tooth crown is compressed antero-posteriorly.

Table 3–1. Morphological types of pharyngeal teeth

Type	General features	Location of tooth hook	Margins	Sub-type	
Type 1	Recurved and conical without grinding surface	At tip			
Type 2	Grinding surface directed backwards or slightly laterad	At anterior end of crown	Lateral and medial	Type 2-0	Grinding surface not concave.
				Type 2-1	Grinding surface concave, usually narrow or small.
				Type 2-2	Grinding surface concave and outspread. Structures such as a U-shaped groove, a bulge, etc., present on grinding surface.
				Type 2-3	Grinding surface concave and narrow, pointing slightly laterad.
Type 3	Grinding surface directed laterad	At medial end of crown	Anterior and posterior	Type 3-0	Grinding surface concave and narrow.
				Type 3-1	Grinding surface outspread.
Type 4	Posterior margin more indistinct than anterior margin	At medial end of crown	Anterior and posterior	Type 4-0	Posterior margin slightly indistinct. Grinding surface usually concave and narrow.
				Type 4-1	Posterior margin significantly indistinct, or grinding surface and posterior side of crown united.
Type 5	Direction of tooth axis making right angle with grinding surface	At medial end of crown	Anterior and posterior	Type 5-0	Shallow groove running along anterior margin on grinding surface.
				Type 5-1	Single or multiple deep grooves running along anterior margin on grinding surface.
				Type 5-2	Tooth crown compressed.

3-1. Danioninae

The pharyngeal dentitions of 59 species in 29 genera of Danioninae were examined. Their teeth are generally arranged in three rows except in *Pararasbora*, *Aphyocypris*, *Parazacco*, and *Gobiocypris*, which have two rows of teeth, and *Esomus* with one row. The typical dental formula is 2(1).4(3).5(4)−5(4).4(3).2(1) in species with three-row dentitions. Ridges and bulges are barely developed on the grinding surface of most species. Although the central teeth of most species, including those with just two rows of teeth, are Type 2−3 or 2−1, Type 2−2 central teeth occur in some species of *Devario* and *Danio*. Exceptionally in this subfamily, the central teeth of *Esomus* are Type 3−1, those of *Apidoparia* are Type 5−2, and those of *Gobiocypris* are Type 4−1 or 4−0. The dentition of *Gobiocypris* corresponds to that of Gobioninae, because the dental formula of both is 2.4−4.2 and their central teeth are Type 4.

***Aaptosyax grypus* Rainboth, 1991** (Fig. 301−1: LBM1210047434, BL 840 mm)

Two specimens from Laos (LBM1210026911 and 1210047434). Dental formula: 2.2.4−4.(2,3).2; all teeth Type 2−1. Tips of tooth crowns sharp. Tooth crowns compressed medio-laterally.

***Raiamas guttatus* (Day, 1870)** (Fig. 301−2: LBM1210014046, BL 105.0 mm)

Four specimens from Thailand (LBM1210014046, 1210033984−1210033986). Dental formula: 2.3.5(4)−5(4).3.2; all teeth Type 2−1. Tips of tooth crowns sharp. Crowns of posterior minor-row teeth bent forward with grinding surfaces oriented dorsad. Crowns of anterior teeth compressed medio-laterally.

***Luciosoma bleekeri* Steindachner, 1878** (Fig. 301−3: LBM1210013835, BL 120.0 mm)

Two specimens from Thailand (LBM1210013834 and 1210013835). Dental formula: 2.4.5−5.4.2; all teeth Type 2−1. Crowns of central teeth and posterior minor-row teeth bent forward with grinding surfaces oriented dorsad. Crowns of anterior teeth compressed medio-laterally.

***Luciosoma setigerum* (Valenciennes, 1842)** (Fig. 301−4: LBM1210013832, BL 125.0 mm)

Three specimens from Thailand (LBM1210013832, 1210013833, and 1210014058). Dental formula: 2(1).4.5(4)−5.4(3).2(1); all teeth Type 2−1. Crowns of central teeth and posterior minor-row teeth bent forward with grinding surfaces oriented dorsad. Crowns of anterior teeth compressed medio-laterally.

***Barilius bendelisis* (Hamilton, 1807)** (Fig. 301−5: LBM1210026921, BL 75.1 mm)

One specimen from Nepal (LBM1210026921). Dental formula: 2.4.4−4.4.2; all teeth Type 2−1. Crowns of central teeth and posterior minor-row teeth bent forward with grinding surfaces oriented dorsad.

***Brevibora dorciocellata* (Duncker, 1904)** (Fig. 301−6: LBM1210015666, BL 18.2 mm)

One specimen from Malaysia (LBM1210015666). Dental formula: 2.3.5−5.4.2; all teeth Type 2−1. Crowns of posterior central and minor-row teeth bent forward with grinding surfaces oriented dorsad.

***Kottelatia brittani* (Axelrod, 1976)** (Fig. 301−7: LBM1210015670, BL 21.4 mm)

One specimen from Indonesia (LBM1210015670). Dental formula: -5.3.1; all teeth Type 2−1. Crowns of posterior central and minor-row teeth bent forward with grinding surfaces oriented dorsad.

***Microrasbora kubotai* (Kottelat and Witte, 1999)** (Fig. 301−8: LBM1210015678, BL 16.4 mm)

One specimen from Thailand (LBM1210015678). Dental formula: 2.3.5−5.3.2, all teeth Type 2−1. Crowns of posterior central and minor-row teeth bent forward with grinding surfaces oriented dorsad.

***Microrasbora rubescens* Annandale, 1918** (Fig. 301−9: LBM1210015676, BL 28.8 mm)

One specimen from Myanmar (LBM1210015676). Dental formula: 2.3.5−5.3.2; central teeth and minor-row teeth Type 2−1, anterior teeth Type 2−1 or 2−0. Crowns of posterior central and minor-row teeth bent forward with grinding surfaces oriented dorsad.

***Opsarius koratensis* (Smith, 1931)** (Fig. 301−10: LBM1210047511, BL 45.2 mm)

Two specimens from Cambodia (LBM1210047510 and 1210047511). Dental formula: (1,2).4.5−5.4.2; central teeth and minor-row teeth Type 2−1, anterior teeth Type 2−1 or 2−0. Crowns of posterior central and minor-row teeth bent forward with grinding surfaces oriented dorsad.

Rasboroides vaterifloris **(Deraniyagala, 1930)** (Fig. 301‒11: LBM1210015674, BL 28.7 mm)

One specimen from Sri Lanka (LBM1210015674). Dental formula: 2.4.5‒5.4.2; all teeth Type 2‒1. Crowns of posterior central teeth and minor-row teeth bent forward with grinding surfaces oriented dorsad.

Salmophasia bacaila **(Hamilton, 1822)** (Fig. 301‒12: LBM1210014082, BL 74.0 mm)

One specimen from Nepal (LBM1210014082). Dental formula: -5.4.1; all teeth Type 2‒1. Crowns of central teeth and posterior minor-row teeth bent forward with grinding surfaces oriented dorsad. Crowns of anterior teeth compressed medio-laterally.

Trigonopoma pauciperforatum **(Weber and de Beaufort, 1916)** (Fig. 301‒13: LBM1210028160, BL 19.2 mm)

Tow specimens from Malaysia (LBM1210015660 and 1210028160). Dental formula: 2.3.5‒5.3.2; all teeth Type 2‒1. Crowns of posterior central and minor-row teeth bent forward with grinding surfaces oriented dorsad.

Rasbora aurotaenia **Tirant, 1885** (Fig. 301‒14: LBM1210047521, BL 54.5 mm)

Two specimens from Cambodia (LBM1210014048 and 1210047521). Dental formula: 2.4.(5,4)‒5.(4,3).2; all teeth Type 2‒1. Crowns of posterior central and minor-row teeth bent forward with grinding surfaces oriented dorsad.

Rasbora borapetensis **Smith, 1934** (Fig. 301‒15: LBM1210015656, BL 22.1 mm)

One specimen from Thailand (LBM1210015656). Dental formula: 1.4.5‒5.4.1; all teeth Type 2‒1. Crowns of central teeth and posterior minor-row teeth bent forward with grinding surfaces oriented dorsad.

Rasbora caudimaculata **Volz, 1903** (Fig. 301‒16: LBM1210026924, BL 77.3 mm)

One specimen from Cambodia (LBM1210026924). Dental formula: 2.4.5‒5.4.2; central teeth Type 2‒3 or 2‒1, anterior teeth and minor-row teeth Type 2‒1. Crowns of posterior central and minor-row teeth bent forward.

Rasbora cephalotaenia **(Bleeker, 1852)** (Fig. 301‒17: LBM1210024424, BL 54.0 mm)

Two specimens from Malaysia (LBM1210024422 and 1210024424). Dental formula: 2.4.5‒5.4.2; central teeth Type 2‒3 or 2‒1, anterior teeth and minor-row teeth Type 2‒1. Crowns of posterior central and minor-row teeth bent forward.

Rasbora daniconius **(Hamilton, 1822)** (Fig. 301‒18: LBM1210047520, BL 45.8 mm)

One specimen from Cambodia (LBM1210047520). Dental formula: 2.4.5‒5.4.2; all teeth Type 2‒1. Crowns of posterior central and minor-row teeth bent forward with grinding surfaces oriented dorsad. Crowns of anterior teeth compressed medio-laterally.

Rasbora dusonensis **(Bleeker, 1850)** (Fig. 301‒19: LBM1210047518, BL 59.0 mm)

One specimen from Cambodia (LBM1210047518). Dental formula: -5.4.2, all teeth Type 2‒1. Crowns of posterior central and minor-row teeth bent forward with grinding surfaces oriented dorsad. Crowns of anterior teeth compressed medio-laterally.

Rasbora einthovenii **(Bleeker, 1851)** (Fig. 301‒20: LBM1210024430, BL 31.8 mm)

One specimen from Malaysia (LBM1210024430). Dental formula: 2.4.5‒5.4.2; all teeth Type 2‒1. Crowns of central teeth and posterior minor-row teeth bent forward with grinding surfaces oriented dorsad. Crowns of anterior teeth compressed medio-laterally.

Rasbora sumatrana **(Bleeker, 1852)** (Fig. 301‒21: LBM1210013993. BL 70.0 mm)

One specimen from Thailand (LBM1210013993). Dental formula: -5.4.2; all teeth Type 2‒1. Crowns of posterior central and minor-row teeth bent forward with grinding surfaces oriented dorsad. Crowns of anterior teeth compressed medio-laterally.

Rasbora trilineata **Steindachner, 1870** (Fig. 301‒22: LBM1210014574, BL 63.2 mm)

Three specimens from Timor-Leste (LBM1210014574 and 1210014575) and Indonesia (LBM1210015658). Dental formula: 2.3.5‒5.4(2).2(1); central teeth and minor-row teeth Type 2‒1, anterior teeth Type 2‒1 or 2‒0. Crowns of posterior central and minor-row teeth bent forward with grinding surfaces oriented dorsad.

Rasbora argyrotaenia (**Bleeker, 1849**) (Fig. 301–23: LBM1210013838, BL 9.8 mm)

One specimen from Thailand (LBM1210013838). Dental formula: 2.4.5–5.4.2; central teeth Type 2–3 or 2–1, anterior teeth and minor-row teeth Type 2–1. Crowns of posterior central teeth and minor-row teeth bent forward.

Rasbora elegans **Volz, 1903** (Fig. 301–24: LBM1210024432, BL 50.0 mm)

One specimen from Malaysia (LBM1210024432). Dental formula: 2.3.5–5.3.2; central teeth Type 2–3 or 2–1, anterior teeth and minor-row teeth Type 2–1. Crowns of posterior central and minor-row teeth bent forward.

Rasbora reticulata **Weber & de Beaufort, 1915** (Fig. 301–25: LBM1210015662, BL 39.1 mm)

Two specimens from Thailand (LBM1210013993 and 1210015662). Dental formula: 2.4.5–5.4.2; central teeth Type 2–3 or 2–1, anterior teeth and minor-row teeth Type 2–1. Crowns of posterior central and minor-row teeth bent forward.

Rasbora tornieri **Ahl, 1922** (Fig. 301–26: LBM1210047516, BL 50.8 mm)

One specimen from Cambodia (LBM1210047516). Dental formula: 2.4.5–5.4.2; central teeth Type 2–3 or 2–1, anterior teeth and minor-row teeth Type 2–1. Crowns of posterior central and minor-row teeth bent forward.

Devario auropurpureus (**Annandale, 1918**) (Fig. 301–27: LBM1210015020, BL 52.3 mm)

One specimen from Myanmar (LBM1210015020). Dental formula: 2.4.5–5.4.1; central teeth Type 2–3 or 2–1, anterior teeth and minor-row teeth Type 2–1. Crowns of posterior central and minor-row teeth bent forward.

Devario devario (**Hamilton, 1822**) (Fig. 301–28: LBM1210026915, BL 44.1 mm)

One specimen from Nepal (LBM1210026915). Dental formula: 2.3.5–5.4.2; all teeth Type 2–1. Crowns of posterior central and minor-row teeth bent forward with grinding surfaces oriented dorsad. Denticles lined up in longitudinal row on grinding surface of posterior major- and minor-row teeth.

Devario aequipinnatus (**McCelland, 1839**) (Fig. 301–29: LBM1210014040, BL 23.4 mm)

Two specimens from Thailand (LBM1210014040

and 1210014041). Dental formula: 2.(3,4).5–5.4.2; central teeth Type 2–2 or 2–1, anterior teeth and minor-row teeth Type 2–1. Crowns of posterior central and minor-row teeth bent forward with grinding surfaces oriented dorsad. Rows of denticles present on grinding surface of posterior major- and minor-row teeth.

Devario malabaricus (**Jerdan, 1849**) (Fig. 301–30: LBM1210014565, BL 58.5 mm)

Five specimens from Thailand (LBM1210014563–1210014565, 1210015033, and 1210015695). Dental formula: 2(1).4(3).5(4)-4(5,3).4(3).2(1); central teeth Type 2–3, anterior teeth and minor-row teeth Type 2–1. Crowns of central teeth and posterior minor-row teeth bent forward. Crowns of anterior teeth compressed medio-laterally.

Danio albolineatus (**Blyth, 1860**) (Fig. 301–31: LBM1210015680, BL 23.7 mm)

One specimen from Thailand (LBM1210015680). Dental formula: 2.2.5–5.3.2; central teeth Type 2–3 or 2–1, anterior teeth and minor-row teeth Type 2–1. Crowns of posterior central and minor-row teeth bent forward.

Danio dangila (**Hamilton, 1822**) (Fig. 301–32: LBM1210015650, BL 56.3 mm)

One specimen from India (LBM1210015650). Dental formula: 2.4.5–5.3.1; all teeth Type 2–1. Crowns of posterior central and minor-row teeth bent forward with grinding surfaces oriented dorsad. Crowns of anterior teeth compressed medio-laterally.

Danio erythoromicron (**Annandale, 1918**) (Fig. 301–33: LBM1210015034, BL 21.0 mm)

One specimen from Myanmar (LBM1210015034). Dental formula: 1.3.5–4.3.1; central teeth Type 2–2, anterior teeth Type 2–1, minor-row teeth Type 2–2 or 2–1. Crowns of posterior central and minor-row teeth bent forward with grinding surfaces orientated dorsad. Longitudinal ridge on grinding surface of posterior major- and minor-row teeth.

Danio kerri **Smith, 1931** (Fig. 301–34: LBM1210015682, BL 31.8 mm)

One specimen from Thailand (LBM1210015682). Dental formula: 2.4.5–5.4.1; central teeth Type 2–2 or 2–1, anterior teeth and minor-row teeth Type 2–1. Crowns of posterior central and minor-row teeth bent forward with grinding surfaces oriented dorsad. Longitudinal ridge on grinding surface of some teeth. Crowns of anterior teeth compressed medio-laterally.

***Danio rerio* (Hamilton, 1822)** (Fig. 301–35: LBM1210015684, BL 31.9 mm)

One specimen from India (LBM1210015684). Dental formula: 2.4.5–5.4.2; central teeth Type 2–2 or 2–1, anterior teeth and minor-row teeth Type 2–1. Crowns of central teeth and posterior minor-row teeth bent forward with grinding surfaces oriented dorsad. Several longitudinal ridges on grinding surface of posterior major- and minor-row teeth.

***Oxygaster anomalura* Van Hasselt, 1823** (Fig. 301–36: LBM1210014073, BL 90.1 mm)

One specimen from Thailand (LBM1210014073). Dental formula: 2.4.4–5.4.2; central teeth Type 2–3 or 2–1, anterior teeth and minor-row teeth Type 2–1. Crowns of posterior central and minor-row teeth bent forward.

***Laubuca caeruleostigmata* Smith, 1931** (Fig. 301–37: LBM1210047480, BL 42.2 mm)

One specimen from Cambodia (LBM1210047480). Dental formula: 2.4.4–5.3.1; central teeth Type 2–3 or 2–1, anterior teeth and minor-row teeth Type 2–1. Crowns of central teeth and posterior minor-row teeth bent forward. Several longitudinal ridges on grinding surface of posterior major- and minor-row teeth.

***Mesobola brevianalis* (Boulenger, 1908)** (Fig. 301–38: LBM1210024364, BL 25.0 mm)

Two specimens from South Africa (LBM1210024362 and 1210024364). Dental formula: 2.3.5–5.3.2; central teeth Type 2–3 or 2–1, anterior teeth and minor-row teeth Type 2–1. Crowns of posterior central and minor-row teeth bent forward. Several denticles present on margin of grinding surface.

***Parachela hypophthalmus* (Bleeker, 1860)** (Fig. 301–39: LBM1210024395, BL 44.0 mm)

Two specimens from Malaysia (LBM1210024393 and 1210024395). Dental formula: 2.4.(4,5)–5.4.2; central teeth Type 2–3 or 2–1, anterior teeth and minor-row teeth Type 2–1. Crowns of posterior central and minor-row teeth bent forward.

***Parachela siamensis* (Günther, 1868)** (Fig. 301–40: LBM1210026940, BL 66.8 mm)

One specimen from Cambodia (LBM1210026940). Dental formula: 2.4.5–4.4.2; central teeth Type 2–3 or 2–1, anterior teeth and minor-row teeth Type 2–1. Crowns of posterior central and minor-row teeth bent forward.

***Parachela williaminae* Fowler, 1934** (Fig. 301–41: LBM1210026941, BL 65.3 mm)

One specimen from Cambodia (LBM1210026941). Dental formula: 2.4.5–5.4.2; central teeth Type 2–3 or 2–1, anterior teeth and minor-row teeth Type 2–1. Crowns of posterior central and minor-row teeth bent forward. Tiny denticles in many longitudinal rows giving wrinkle-like appearance to each grinding surface.

***Candidia barbata* (Regan, 1908)** (Fig. 301–42: LBM1210047477, BL 47.7 mm)

One specimen from Taiwan, China (LBM1210047477). Dental formula: 2.4.5–5.4.2; all teeth Type 2–1. Crowns of posterior central and minor-row teeth bent forward with grinding surfaces oriented dorsad.

***Opsariichthys bidens* Günther, 1873** (Fig. 301–43: LBM1210014570, BL 117.6 mm)

Five specimens from Hubei, China (LBM1210014082 and 1210014570–1210014573). Dental formula: 1.3.4–5(4).3(4).1; central teeth Type 2–3 or 2–1, anterior teeth Type 2–1 or 2–0, minor-row teeth Type 2–1. Crowns of posterior central and minor-row teeth bent forward.

***Opsariichthys uncirostris* (Temminck and Schlegel, 1846)** (Fig. 301–44: LBM1210013774, BL 169.0 mm)

19 specimens from Shiga, Japan (LBM1210003627, 1210013614, 1210013640, 1210013774, 1210013775, 1210013818, 1210013819, 1210013936, 1210013937, 1210014396–1210014399, 1210014597–1210014599, 1210039862, 1210039863, and 1210040517). Dental formula: 2(1).4(3).5(4)–4(5).4(3).2(1); central teeth Type 2–3 or 2–1, anterior teeth Type 2–1 or 2–0, minor-row teeth Type 2–1. Crowns of central teeth and posterior minor-row teeth bent forward.

***Opsariichthys pachycephalus* Günther, 1868** (Fig. 301–45: LBM1210014593, BL 94.0 mm)

Two specimens from Taiwan, China (LBM1210014593 and 1210014594). Dental formula: 2.4.5–4.4.2; central teeth Type 2–3 or 2–1, anterior teeth

Type 2-1 or 2-0, minor-row teeth Type 2-1. Crowns of posterior central and minor-row teeth bent forward.

***Zacco platypus* (Temminck and Schlegel, 1846)** (Fig. 301-46: LBM1210014449, BL 92.0 mm)

Eleven specimens from Shiga and Gifu, Japan (LBM1210003606-1210003608, 1210013563, 1210013637, 1210013638, 1210013644, 1210013779, 1210013780, 1210013865, 1210013938, 1210013939, 1210014447-1210014450, 1210039914, and 1210043937). Dental formula: 2(1).4(3).5(4)-(4,5).4(3).2(1); central teeth Type 2-3 or 2-1, anterior teeth Type 2-1 or 2-0, minor-row teeth Type 2-1. Crowns of posterior central and minor-row teeth bent forward.

***Nipponocypris sieboldii* (Temminck and Schlegel, 1846)** (Fig. 301-47: LBM1210024517, BL 92.6 mm)

Eleven specimens from Shiga, Japan (LBM1210024517, 1210024521, 1210024523, 1210024525, 1210028154, 1210028156, 1210039858-1210039860, 1210046450, and 1210046452). Dental formula: 1(2).4.5(4)-4(5).4(3).1(2); central teeth Type 2-3 or 2-1, anterior teeth Type 2-1 or 2-0, minor-row teeth Type 2-1. Crowns of posterior central and minor-row teeth bent forward.

***Nipponocypris temminckii* (Temminck and Schlegel, 1846)** (Fig. 301-48: LBM1210039895, BL 110.0 mm)

Eighteen specimens from Shiga, Japan (LBM1210003609-1210003614, 1210013772, 1210013778, 1210013950, 1210014668, 1210014669, 1210024467, 1210024519, 1210039893, 1210039895, 1210040071, 1210043019, and 1210048836). Dental formula: 2(1).4.5(4)-4.4.2(1); central teeth Type 2-3 or 2-1, anterior teeth Type 2-1 or 2-0, minor-row teeth Type 2-1. Crowns of posterior central and minor-row teeth bent forward.

***Pararasbora moltrechti* Regan, 1908** (Fig. 301-49: LBM1210036857, BL 46.8 mm)

One specimen from Taiwan, China (LBM1210036857). Dental formula: 2.4-4.3; all teeth Type 2-1. Crowns of posterior central teeth and minor-row teeth bent forward with grinding surfaces oriented dorsad.

***Aphyocypris kikuchii* (Oshima, 1919)** (Fig. 301-50: LBM1210036440, BL 69.5 mm)

Ten specimens from Taiwan, China (LBM1210036431-1210036440). Dental formula: 3(2,4).4(5)-4(5).4(2,3); central teeth Type 2-3 or 2-1, anterior teeth Type 2-1 or 2-0, and minor-row teeth Type 2-1. Crowns of posterior central and minor-row teeth bent forward.

***Aphyocypris chinensis* Günther, 1868** (Fig. 301-51: LBM1210026123, BL 60.0 mm)

71 specimens from Fukuoka, Japan, (LBM1210026123, 1210015026, 1210036375-1210036417, 1210036419-1210036793). Dental formula: 3(2).(4,5)-(4,5).3(2); central teeth Type 2-3 or 2-1, anterior teeth Type 2-1 or 2-0, minor-row teeth Type 2-1. Crowns of posterior central and minor-row teeth bent forward.

***Parazacco spilurus* (Günther, 1868)** (Fig. 301-52: IHCAS837395, BL 73.8 mm)

One specimen from Guangxi, China (IHCAS837395). Dental formula: 3.4-5.3; central teeth Type 2-3 or 2-1, anterior teeth Type 2-0, minor-row teeth Type 2-1. Crowns of posterior central and minor-row teeth bent forward.

***Hemigrammocypris rasborella* Fowler, 1910** (Fig. 301-53: LBM1210015004, BL 55.8 mm)

27 specimens from Shiga, Japan (LBM1210015003, 1210015004, 1210037064, 1210036425-1210036427, 1210036441-1210036450, 1210040439, 1210040441, 1210040443, 1210040445, 1210040447, 1210040449, 1210042009, 1210042011, 1210042013, 1210042266, 1210042268, 1210042270). Dental formula: (1,2).(3,4).(4,5)-5(4).4(3).(1,2); central teeth and minor-row teeth Type 2-3 or 2-1, anterior teeth Type 2-1 or 2-0. Crowns of two most posterior central teeth bent forward. Crowns of posterior central and minor-row teeth bent forward. Hamulus grooves present on the posterior or medial side of crown of posterior major- and minor-row teeth.

***Esomus caudiocellatus* Ahl, 1923** (Fig. 301-54: LBM1210014988, BL 34.6 mm)

One specimen from Indonesia (LBM1210014988). Dental formula: 5-5; central teeth Type 3-1, and anterior teeth Type 3-1 or 2-0. Crowns of some teeth worn.

***Esomus metallicus* Ahl, 1923** (Fig. 301-55: LBM1210026929 , BL 41.9 mm)

One specimen from Thailand (LBM1210026929). Dental formula: 4-5; central teeth Type 3-1, anterior teeth

Type 3-1 or 2-0. Crowns of some teeth worn.

***Esomus danricus* (Hamilton, 1822)** (Fig. 301-56: LBM1210014081, BL 66.0 mm)

Two specimens from Nepal (LBM1210014081 and 1210014084). Dental formula: (4,5)-4; central teeth Type 3-1, anterior teeth Type 3-0. Crowns of some teeth worn.

***Esomus longimanus* (Lunel, 1881)** (Fig. 301-57: LBM1210026932, BL 54.6 mm)

One specimen from Cambodia (LBM1210026932). Dental formula: 4-4; all teeth Type 3-1. Crowns of some teeth worn.

***Aspidoparia morar* (Hamilton, 1822)** (Fig. 301-58 : LBM1210015741, BL 68.5 mm)

Two specimens from Nepal (LBM1210013996 and 1210015741). Dental formula: 2.3.4-4.3.2; central teeth Type 5-2, anterior teeth Type 2-2, minor-row teeth Type 2-3 or 2-1. Crowns of posterior minor-row teeth bent forward. Crowns of central teeth compressed antero-posteriorly. Minor-row teeth slender and rod-like, major-row teeth worn.

***Gobiocypris rarus* Ye and Fu, 1983** (Fig. 301-59: LBM1210037063, BL 34.4 mm)

Five specimens from China (LBM1210037061- 1210037065). Dental formula: 2.4-4.2; central teeth Type 4-1 or 4-0, anterior teeth and minor-row teeth Type 2-1. Crowns of posterior central and minor-row teeth bent forward. Several denticles lining anterior margins of central teeth.

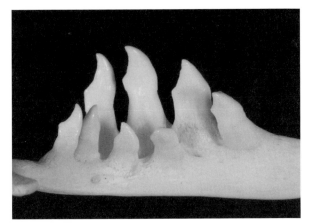

Fig. 301-1. *Aaptosyax grypus*: LBM1210047434.

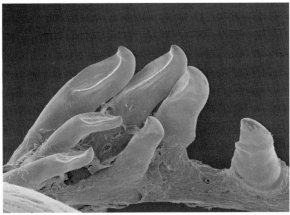

Fig. 301-2. *Raiamas guttatus*: LBM1210014046.

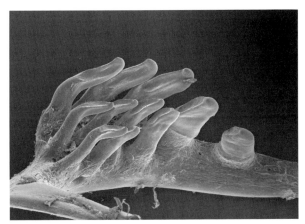

Fig. 301-3. *Luciosoma bleekeri*: LBM1210013835.

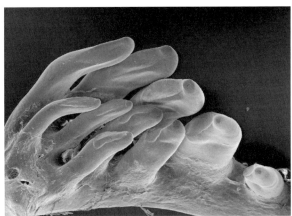

Fig. 301-4. *Luciosoma setigerum*: LBM1210013832.

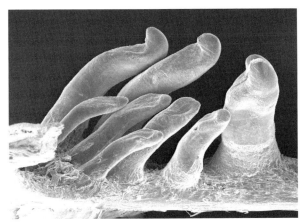

Fig. 301-5. *Barilius bendelisis*: LBM1210026921.

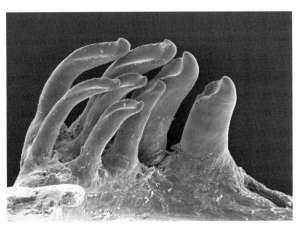

Fig. 301-6. *Brevibora dorciocellata*: LBM1210015666.

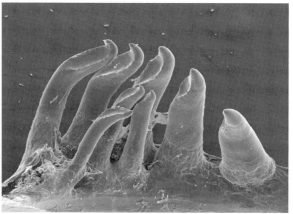

Fig. 301-7. *Kottelatia brittani*: LBM1210015670.

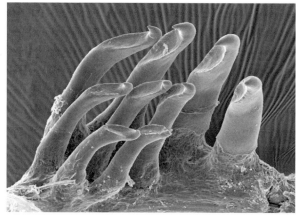

Fig. 301-8. *Microrasbora kubotai*: LBM1210015678.

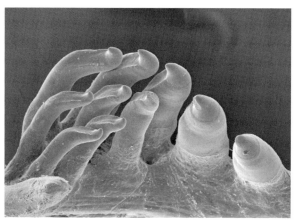

Fig. 301-9. *Microrasbora rubescens*: LBM1210015676.

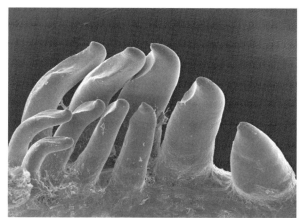

Fig. 301-10. *Opsarius koratensis*: LBM1210047511.

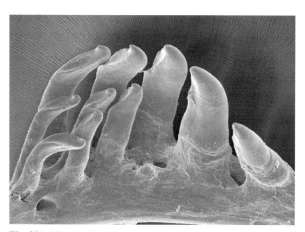

Fig. 301-11. *Rasboroides vaterifloris*: LBM1210015674.

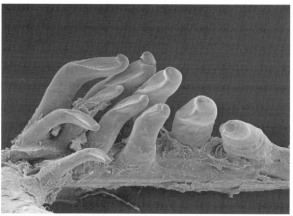

Fig. 301-12. *Salmophasia bacaila*: LBM1210014082.

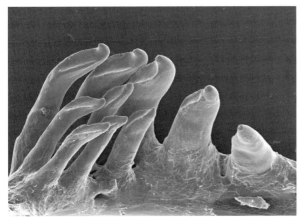

Fig. 301-13. *Trigonopoma pauciperforatum*: LBM1210028160.

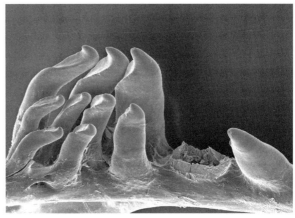

Fig. 301-14. *Rasbora aurotaenia*: LBM1210047521.

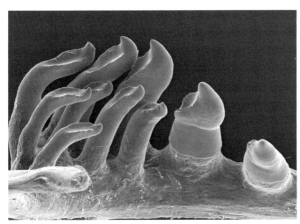

Fig. 301-15. *Rasbora borapetensis*: LBM1210015656.

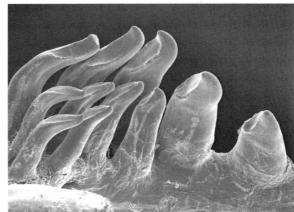

Fig. 301-16. *Rasbora caudimaculata*: LBM1210026924.

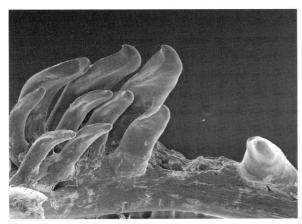

Fig. 301-17. *Rasbora cephalotaenia*: LBM1210024424.

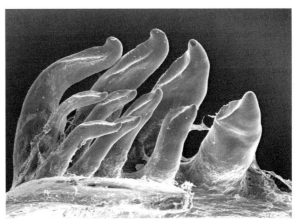

Fig. 301-18. *Rasbora daniconius*: LBM1210047520.

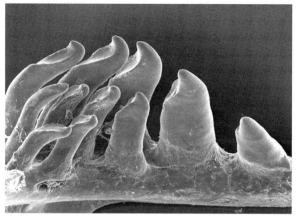

Fig. 301-19. *Rasbora dusonensis*: LBM1210047518.

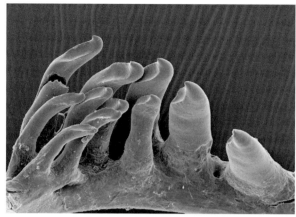

Fig. 301-20. *Rasbora einthovenii*: LBM1210024430.

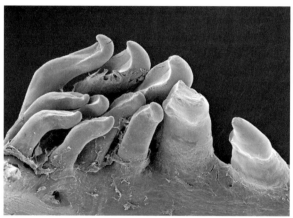

Fig. 301-21. *Rasbora sumatrana*: LBM1210013993.

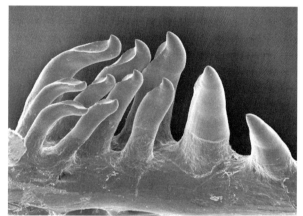

Fig. 301-22. *Rasbora trilineata*: LBM1210014574.

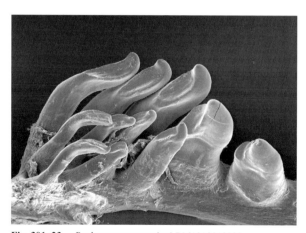

Fig. 301-23. *Rasbora argyrotaenia*: LBM1210013838.

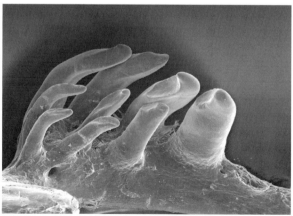

Fig. 301-24. *Rasbora elegans*: LBM1210024432.

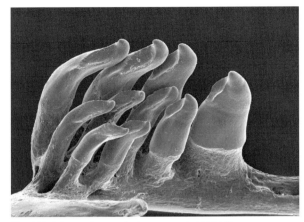
Fig. 301-25. *Rasbora reticulata*: LBM1210015662.

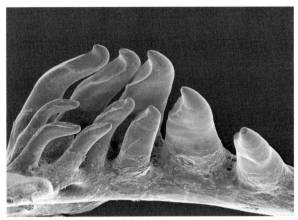
Fig. 301-26. *Rasbora tornieri*: LBM1210047516.

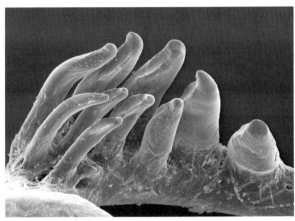
Fig. 301-27. *Devario auropurpureus*: LBM1210015020.

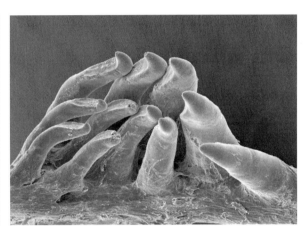
Fig. 301-28. *Devario devario*: LBM1210026915.

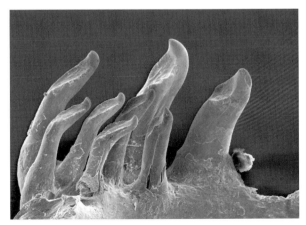
Fig. 301-29. *Devario aequipinnatus*: LBM1210014040.

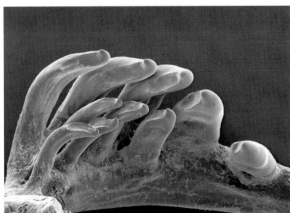
Fig. 301-30. *Devario malabaricus*: LBM1210014565.

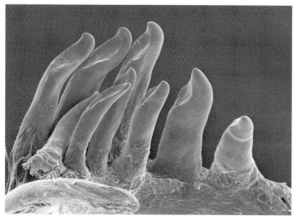

Fig. 301-31. *Danio albolineatus*: LBM1210015680.

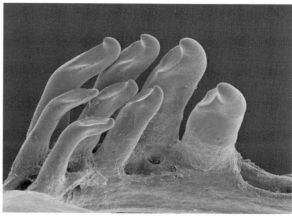

Fig. 301-32. *Danio dangila*: LBM1210015650.

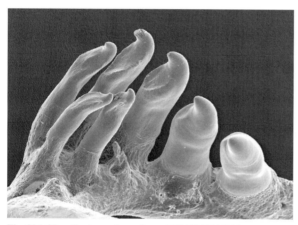

Fig. 301-33. *Danio erythoromicron*: LBM1210015034.

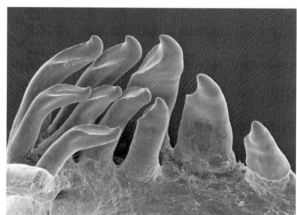

Fig. 301-34. *Danio kerri*: LBM1210015682.

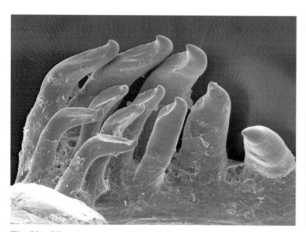

Fig. 301-35. *Danio rerio*: LBM1210015684.

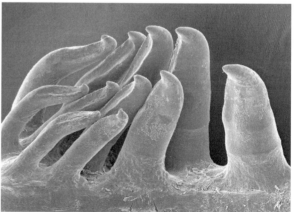

Fig. 301-36. *Oxygaster anomalura*: LBM1210014073.

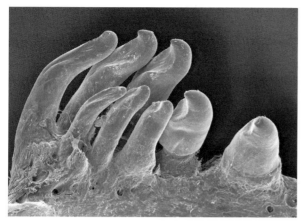

Fig. 301-37. *Laubuca caeruleostigmata*: LBM1210047480.

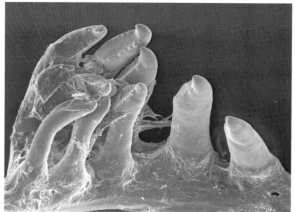

Fig. 301-38. *Mesobola brevianalis*: LBM1210024364.

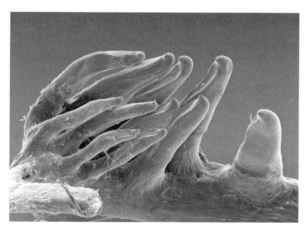

Fig. 301-39. *Parachela hypophthalmus*: LBM1210024395.

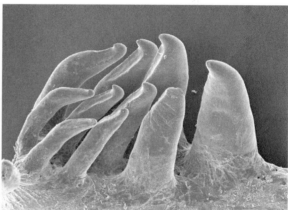

Fig. 301-40. *Parachela siamensis*: LBM1210026940.

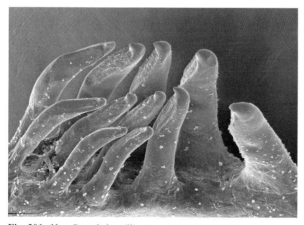

Fig. 301-41. *Parachela williaminae*: LBM1210026941.

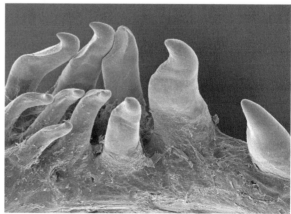

Fig. 301-42. *Candidia barbata*: LBM1210047477.

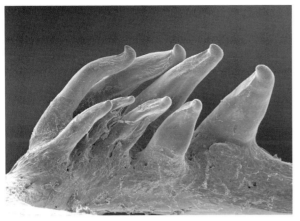

Fig. 301-43. *Opsariichthys bidens*: LBM1210014570.

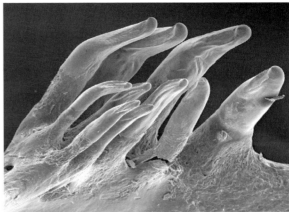

Fig. 301-44. *Opsariichthys uncirostris*: LBM1210013774.

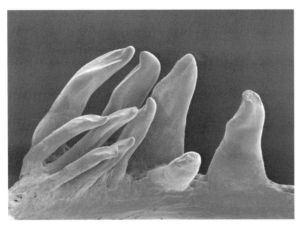

Fig. 301-45. *Opsariichthys pachycephalus*: LBM1210014593.

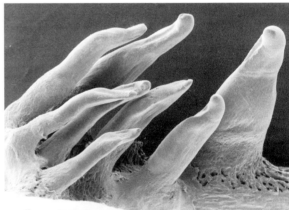

Fig. 301-46. *Zacco platypus*: LBM1210014449.

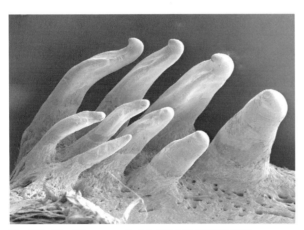

Fig. 301-47. *Nipponocypris sieboldii*: LBM1210024517.

Fig. 301-48. *Nipponocypris temminckii*: LBM1210039895.

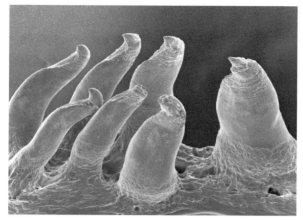

Fig. 301-49. *Pararasbora moltrechti*: LBM1210036857.

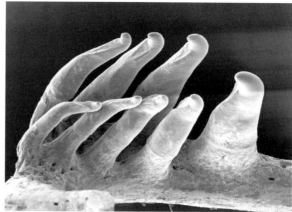

Fig. 301-50. *Aphyocypris kikuchii*: LBM1210036440.

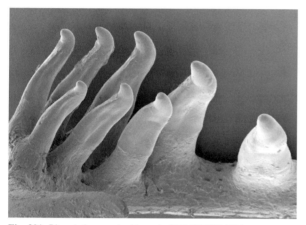

Fig. 301-51. *Aphyocypris chinensis*: LBM1210026123.

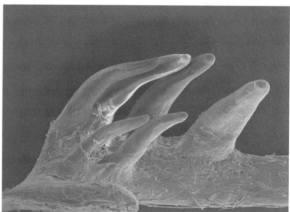

Fig. 301-52. *Parazacco spilurus*: IHCAS837395.

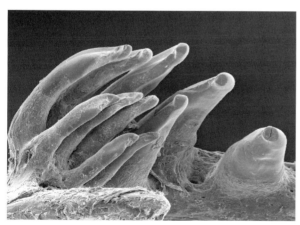

Fig. 301-53. *Hemigrammocypris rasborella*: LBM1210015004.

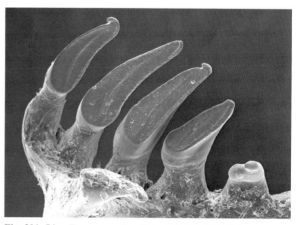

Fig. 301-54. *Esomus caudiocellatus*: LBM1210014988.

3. Descriptions of the pharyngeal dentition of cyprinid subfamilies　43

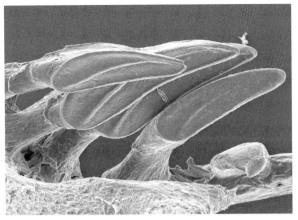

Fig. 301-55. *Esomus metallicus*: LBM1210026929.

Fig. 301-56. *Esomus danricus*: LBM1210014081.

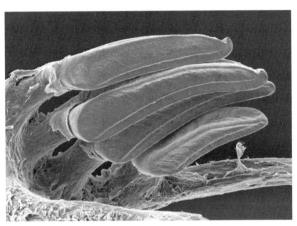

Fig. 301-57. *Esomus longimanus*: LBM1210026932.

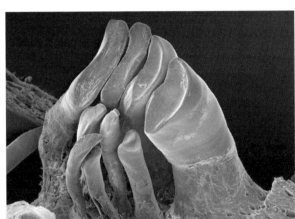

Fig. 301-58. *Aspidoparia morar*: LBM1210015741.

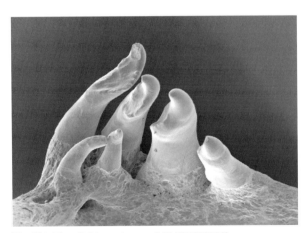

Fig. 301-59. *Gobiocypris rarus*: LBM1210037063.

3-2. Barbinae

The pharyngeal dentitions of 102 species in 31 genera of Barbinae were examined. Their teeth are always arranged in three rows and the typical dental formula is 2(1).3(2).5(4)−5(4).3(2).2(1), except for the one-row dentition of *Catlocarpio* with the dental formula 4−4. The number of teeth in each row in Barbinae is similar to that in Danioninae, but there is one tooth fewer in the B row. The central teeth are usually Type 2−2. In contrast to the Danioninae, no central teeth are Type 2−3. The tooth at position An3 (tooth A1) is small and Type 2−1 or 2−0 in many species, but the tooth at position An2 (tooth A1 or A2) is usually Type 2−2 or 2−1. *Catlocarpio* has a quite different dentition, in which the teeth are arranged in a single row and the central teeth are Type 4−1 or 5−0.

Hampala dispar **Smith, 1934** (Fig. 302−1: LBM1210013989, BL 123.0 mm)

One specimen from Thailand (LBM1210013989). Dental formula: 1.3.5−5.3.1; all teeth Type 2−1. Crowns of posterior central and minor-row teeth bent forward with grinding surfaces oriented dorsad. Tips of tooth crowns sharp. Grinding surfaces of posterior major- and minor-row teeth wrinkled. Tooth crown of anterior teeth compressed medio-laterally.

Hampala macrolepidota **Kuhl and Van Hasselt, 1823** (Fig. 302−2: LBM1210013990, BL 145.0 mm)

Four specimens from Thailand (LBM1210013990, 1210013991, 1210014036, and 1210014037). Dental formula: 1.3.4−5(4).3.1; all teeth Type 2−1. Crowns of posteror minor-row teeth bent forward with grinding surfaces oriented dorsad. Tips of tooth crowns sharp. Grinding surfaces of posterior major- and minor-row teeth wrinkled. Crowns of anterior teeth compressed mediolaterally.

Percocypris pingi (**Tchang, 1930**) (Fig. 302−3: IHCAS8180235, BL 218.0 mm)

Two specimens from Hubei, China (IHCAS8180235). Dental formula: 2.3.4−5.3.1; all teeth Type 2−1. Crowns of posterior central and minor-row teeth bent forward with grinding surfaces oriented dorsad.

Percocypris regani (**Tchang, 1935**) (Fig. 302−4: IHCAS81104166, BL 106.0 mm)

One specimen from Yunnan, China (IHCAS81104166).

Dental formula: 2.3.5−5.3.2; all teeth Type 2−1. Crowns of posterior central and minor-row teeth bent forward with grinding surfaces oriented dorsad.

Acrossocheilus beijiangensis **Wu & Lin, 1977** (Fig. 302−5: IHCAS874480, BL 120.0 mm)

One specimen from Guangxi, China (IHCAS874480). Dental formula: 2.3.5−5.3.2; central teeth and minor-row teeth Type 2−2 or 2−1, anterior teeth Type 2−1 or 2−0. Crowns of posterior central and minor-row teeth bent forward with grinding surfaces oriented dorsad.

Acrossocheilus fasciatus (**Steindachner, 1892**) (Fig. 302−6: IHCAS59122, BL 109.0 mm)

One specimen from Zhejiang, China (IHCAS59122) Dental formula: 2.3.5−5.3.2; central teeth Type 2−2, anterior teeth Type 2−1 or 2−0, minor-row teeth Type 2−2. Crowns of posterior central and minor-row teeth bent forward with grinding surfaces oriented dorsad. Both margins of grinding surfaces of central teeth and minor-row teeth jutting out.

Acrossocheilus iridescens (**Nichols & Pope, 1927**) (Fig. 302−7: IHCAS602243, BL 128.0 mm)

One specimen from Guangdong, China (IHCAS602243). Dental formula: 2.3.5−5.3.2; central teeth and minor-row teeth Type 2−2, anterior teeth Type 2−1 or 2−0. Crowns of posterior central and minor-row teeth bent forward with grinding surfaces oriented dorsad. Both margins of grinding surfaces of central teeth and minor-row teeth jutting out.

Acrossocheilus kreyenbergii (**Regan, 1908**) (Fig. 302−8: IHCAS839232, BL 46.5 mm)

Two specimens from Sichuan, China (IHCAS849090 and 839232). Dental formula: 2.3.(4,5)−(4,5).3.2; central teeth and minor-row teeth Type 2−2, anterior teeth Type 2−1 or 2−0. Crowns of posterior central and minor-row teeth bent forward with grinding surfaces oriented dorsad. Longitudinal ridge on grinding surface of posterior major- and minor-row teeth.

Acrossocheilus longipinnis (**Wu, 1939**) (Fig. 302−9: IHCAS875927, BL 162.0 mm)

One specimen from Guizhou, China (IHCAS875927). Dental formula: -5.3.2; central teeth and minor-row teeth Type 2−2, anterior teeth Type 2−1 or 2−0. Crowns of posterior central and minor-row teeth bent forward with grinding surfaces oriented dorsad. Longitudinal ridge on grinding surface of posterior major- and minor-row teeth.

***Acrossocheilus wenchowensis* Wang, 1935** (Fig. 302-10: IHCAS7461127, BL 112.0 mm)

One specimen from Fujian, China (IHCAS7461127). Dental formula: 2.3.5-5.3.1; central teeth Type 2-2, anterior teeth Type 2-1 or 2-0, minor-row teeth Type 2-2 or 2-1. Crowns of posterior central and minor-row teeth bent forward with grinding surfaces oriented dorsad. Longitudinal ridge on grinding surface of posterior major- and minor-row teeth.

***Acrossocheilus hemispinus* (Nichols, 1925)** (Fig. 302-11: IHCAS81104633, BL 117.0 mm)

One specimen from Guangxi, China (IHCAS81104633). Dental formula is 2.3.5-4.3.2; central teeth and minor-row teeth Type 2-2, anterior teeth Type 2-1 or 2-0. Crowns of posterior central and minor-row teeth bent forward with grinding surfaces oriented dorsad. Several longitudinal ridges on grinding surface of central teeth. Both margins of grinding surface of central teeth and anterior minor-row teeth jutting out.

***Acrossocheilus monticola* (Günther, 1888)** (Fig. 302-12: IHCAS8061243, BL 111.0 mm)

One specimen from Shanxi, China (IHCAS8061243). Dental formula: 2.3.5-4.3.2; central teeth and minor-row teeth Type 2-2. Crowns of central teeth and posterior minor-row teeth bent forward with grinding surfaces oriented dorsad. U-shaped groove on grinding surface of central teeth and posterior minor-row teeth.

***Acrossocheilus paradoxus* (Günther, 1868)** (Fig. 302-13: LBM1210024283, BL 50.0 mm)

Ten specimens from Taiwan, China (LBM1210013851-1210013854, 1210014957, 1210014958, 1210024281, 1210024283, 1210047479) and Zhejiang, China (IHCAS7494547). Dental formula: 2(1).3.5(4)-5(4).3(2).2(1); central teeth and minor-row teeth Type 2-2, anterior teeth Type 2-1 or 2-0. Crowns of central teeth and posterior minor-row teeth bent forward with grinding surfaces oriented dorsad. Both margins of grinding surface of central teeth and minor-row teeth jutting out.

***Acrossocheilus parallens* (Nichols, 1931)** (Fig. 302-14: IHCAS9051943, BL 126.0 mm)

One specimen from Jiangxi, China (IHCAS9051943). Dental formula: 2.3.5-5.3.2; central teeth and minor-row teeth Type 2-2, anterior teeth Type 2-1 or 2-0. Crowns of central teeth and posterior minor-row teeth bent forward with grinding surfaces oriented dorsad. Both margins of grinding surface of posterior major- and minor-row teeth jutting out.

***Acrossocheilus yunnanensis* (Regan, 1904)** (Fig. 302-15: IHCAS8180252, BL 138.0 mm)

Two specimens from Hubei (IHCAS8180252) and Sichuan, China (IHCAS585071). Dental formula: (1,2).3.5-5.3.2; central teeth and minor-row teeth Type 2-2, anterior teeth Type 2-1 or 2-0. Crowns of central teeth and posterior minor-row teeth bent forward with grinding surfaces oriented dorsad. Both margins of grinding surface of posterior major- and minor-row teeth jutting out.

***Anematichthys armatus* (Valenciennes, 1842)** (Fig. 302-16: LBM1210026904, BL 74.4 mm)

Two specimens from Cambodia (LBM1210026904 and 1210026906). Dental formula: (1,2).(2,3).4-4.3.2; central teeth and minor-row teeth Type 2-2, anterior teeth Type 2-1 or 2-0. Crowns of posterior central and minor-row teeth bent forward with grinding surfaces oriented dorsad. Both margins of grinding surface of central teeth and minor-row teeth jutting out.

***Anematichthys repasson* (Bleeker, 1853)** (Fig. 302-17: LBM1210013849, BL 105.4 mm)

Two specimens from Thailand (LBM1210013849 and 1210013450). Dental formula: 2.3.5-5.3.2(0,1.); central teeth and minor-row teeth Type 2-2, anterior teeth Type 2-1. Crowns of posterior central and minor-row teeth bent forward with grinding surfaces oriented dorsad. Both margins of grinding surface of central teeth and minor-row teeth jutting out. Crowns of anterior teeth compressed medio-laterally.

***Balantiocheilos melanopterus* (Bleeker, 1850)** (Fig. 302-18: LBM1210014973, BL 57.3 mm)

Five specimens from Cambodia (LBM1210014972. 1210014973, 1210014974, 1210014975, and 1210026124). Dental formula: 2(1).3(2).5-5.3(2).2(1); central teeth and minor-row teeth Type 2-2, anterior teeth

Type 2-2 or 2-1. Crowns of central teeth and posterior minor-row teeth bent forward with grinding surfaces oriented dorsad.

Barbodes semifasciolatus (Günther, 1868) (Fig. 302-19: IHCAS8243308, BL 48.5 mm)

Two specimens from Hainan, China (LBM1210015720 and IHCAS8243308). Dental formula: 2.3.5-5.3.2; central teeth Type 2-2 or 2-1, anterior teeth Type 2-1 or 2-0, minor-row teeth Type 2-2. Crowns of central teeth and posterior minor-row teeth bent forward with grinding surfaces oriented dorsad.

Barbodes wynaadensis (Day, 1873) (Fig. 302-20: IHCAS58530, BL 174.0 mm)

One specimen from Yunnan, China (IHCAS58530). Dental formula: 2.3.5-5.3.2; central teeth and minor-row teeth Type 2-2, anterior teeth Type 2-1 or 2-0. Crowns of posterior central teeth and minor-row teeth bent forward with grinding surfaces oriented dorsad. U-shaped groove and knobby bulge on grinding surface of central teeth.

Barbonymus altus (Günther, 1868) (Fig. 302-21: LBM1210026933, BL 70.2 mm)

One specimen from Thailand (LBM1210026933). Dental formula: 2.3.5-4.3.1; central teeth Type 2-2, anterior teeth Type 2-1, minor-row teeth Type 2-2 or 2-1. Crowns of posterior central teeth and minor-row teeth bent forward with grinding surfaces oriented dorsad. U-shaped groove on grinding surface of posterior central teeth and minor-row teeth.

Barbonymus gonionotus (Bleeker, 1849) (Fig. 302-22: LBM1210047432, BL 165.0 mm)

Two specimens from Thailand (LBM1210013994 and 1210047432). Dental formula: 2.3.5-(4,5).3.(1,2); central teeth and minor-row teeth Type 2-2, anterior teeth Type 2-1. Crowns of posterior central and minor-row teeth bent forward with grinding surfaces oriented dorsad. U-shaped groove on grinding surface of posterior central and minor-row teeth.

Barbus bifrenatus Fowler, 1935 (Fig. 302-23: LBM1210024287, BL 30.5 mm)

One specimen from Botswana (LBM1210024287). Dental formula: 1.3.4-4.3.2; central teeth and minor-row teeth Type 2-2, anterior teeth Type 2-1. Crowns of central teeth and posterior minor-row teeth bent forward with

grinding surfaces oriented dorsad.

Barbus tauricus Kessler, 1877 (Fig. 302-24: LBM1210013972, BL 187.1 mm)

One specimen from European Russia (LBM1210013972). Dental formula: -4.3.1; central teeth and minor-row teeth Type 2-2 or 2-1, anterior teeth Type 2-0. Crowns of posterior central and minor-row teeth bent forward with grinding surfaces oriented dorsad.

Barbus afrohamiltoni Crass, 1960 (Fig. 302-25: LBM1210024285, BL 61.0 mm)

One specimen from South Africa (LBM1210024285). Dental formula: 2.3.5-5.3.2; central teeth and minor-row teeth Type 2-2, anterior teeth Type 2-1 or 2-0. Crowns of posterior central and minor-row teeth bent forward with grinding surfaces oriented dorsad. Tiny denticles in longitudinal rows looking like wrinkles on grinding surface of central teeth and minor-row teeth.

Barbus barbus (Linnaeus, 1758) (Fig. 302-26: LBM1210013971, BL 158 mm)

One specimen from Europian Russia (LBM1210013971). Dental formula: -4.3.1; central teeth and minor-row teeth Type 2-2, anterior teeth Type 2-0. Crowns of central teeth and posterior minor-row teeth bent forward with grinding surfaces oriented dorsad.

Barbus bynni (Forsskål, 1775) (Fig. 302-27: LBM1210014045, BL 195.0 mm)

Two specimens from Egypt (LBM1210014045). Dental formula: 2.3.5-4.3.2; central teeth and minor-row teeth Type 2-2, anterior teeth Type 2-1. Crowns of posterior central teeth and minor-row teeth well worn and bent forward with grinding surfaces oriented dorsad. U-shaped groove on grinding surface of posterior major- and minor-row teeth.

Barbus jae Boulenger, 1903 (Fig. 302-28: LBM1210047615, BL 20.9 mm)

One specimen from Cameroon (LBM1210047615). Dental formula: 2.2.5-5.2.1; central teeth and minor-row teeth Type 2-2, anterior teeth Type 2-2 or 2-1. Crowns of central teeth and minor-row teeth bent forward with grinding surfaces oriented dorsad. Three longitudinal rows of denticles on grinding surface of all teeth.

Barbus paludinosus **Peters, 1852** (Fig. 302–29:
LBM1210024291, BL 36.0 mm)

Two specimens from Angola (1210024289 and
1210024291). Dental formula: 2.3.5–(4,5).3.2; central
teeth and minor-row teeth Type 2–2, anterior teeth Type
2–1 or 2–0. Crowns of central teeth and posterior minor-
row teeth bent forward with grinding surfaces oriented
dorsad.

Barbus peloponnesius **Valenciennes, 1842** (Fig. 302–
30: LBM1210013842, BL 97.0 mm)

Two specimens from Romania (LBM1210013842
and 1210013859). Dental formula: 2.3.5–5.(2,3).2; central
teeth and minor-row teeth Type 2–2, anterior teeth Type
2–2 or 2–1. Crowns of central teeth and posterior minor-
row teeth bent forward with grinding surfaces oriented
dorsad. Both margins of grinding surface of posterior
central and minor-row teeth jutting out.

Barbus toppini **Boulenger, 1916** (Fig. 302–31:
LBM1210024293, BL 25.0 mm)

One specimen from Malawi (LBM1210024293).
Dental formula: 2.3.5–5.3.2; central teeth and minor-row
teeth Type 2–2, anterior teeth Type 2–1 or 2–0. Crowns of
central teeth and posterior minor-row teeth bent forward
with grinding surfaces oriented dorsad. Several denticles
present on margins of grinding surface.

Barbus trimaculatus **Peters, 1852** (Fig. 302–32:
LBM1210024294, BL 63.4 mm)

One specimen from Malawi (LBM1210024294).
Dental formula: 2.3.5–5.3.2; central teeth and minor-row
teeth Type 2–2, anterior teeth Type 2–2 or 2–0. Crowns of
posterior central and minor-row teeth bent forward with
grinding surfaces oriented dorsad. Several fine parallel
longitudinal ridges on grinding surface of posterior major-
and minor-row teeth.

Cosmochilus harmandi **Sauvage, 1878** (Fig. 302–33:
LBM1210047608, BL 110.0 mm)

One specimen from Cambodia (LBM1210047608).
Dental formula: 2.3.5–5.3.2; central teeth and minor-
row teeth Type 2–2, anterior teeth Type 2–0. Crowns of
posterior central and minor-row teeth bent forward with
grinding surfaces oriented dorsad. Transverse groove on
grinding surface of posterior major- and minor-row teeth.
Tooth A2 massive.

Cyclocheilichthys apogon **(Valenciennes, 1842)** (Fig.
302–34: LBM1210013848, BL 57.2 mm)

Three specimens from Thailand (LBM1210013839,
1210013848, and 1210014063). Dental formula: 2.3.4–
4.3.2; central teeth and minor-row teeth Type 2–2, anterior
teeth Type 2–1 or 2–0. Crowns of posterior central and
minor-row teeth bent forward with grinding surfaces
oriented dorsad. Crowns of anterior teeth compressed.

Cyclocheilichthys enoplos **(Bleeker, 1849)** (Fig. 302–35:
LBM1210014064, BL 229.0 mm)

One specimen from Thailand (LBM1210014064).
Dental formula: 2.3.5–5.3.2; central teeth and minor-
row teeth Type 2–2, anterior teeth Type 2–0. Crowns of
posterior central and minor-row teeth bent forward with
grinding surfaces oriented dorsad. U-shaped groove on
grinding surface of posterior major- and minor-row teeth.
Tooth A2 massive.

Cyclocheilichthys furcatus **Sontirat, 1989** (Fig. 302–36:
LBM1210047486, BL 117.0 mm)

One specimen from Cambodia (LBM1210047486).
Dental formula: 2.3.5–5.3.2; central teeth and minor-row
teeth Type 2–2, anterior teeth Type 2–1 or 2–0. Crowns of
posterior central and minor-row teeth bent forward with
grinding surfaces oriented dorsad. U-shaped groove on
grinding surface of posterior major- and minor-row teeth.
Tooth A2 massive.

Cyclocheilichthys lagleri **Sontirat, 1989** (Fig. 302–37:
LBM1210026943, BL 53.7 mm)

One specimen from Cambodia (LBM1210026943).
Dental formula: 2.2.4–4.2.2; central teeth and minor-row
teeth Type 2–2, anterior teeth Type 2–1 or 2–0. Crowns of
posterior central and minor-row teeth bent forward with
grinding surfaces oriented dorsad. Crowns of anterior teeth
compressed.

Cyprinion kais **Heckel, 1843** (Fig. 302–38: LBM1210013984,
BL 87.0 mm)

One specimen from Iraq (LBM1210013984). Dental
formula: –4.3.2; central teeth and minor-row teeth Type
2–2. Crowns of posterior central and minor-row teeth bent
forward with grinding surfaces oriented dorsad. Some
longitudinal ridges and U-shaped groove on grinding
surface of posterior major- and minor-row teeth.

***Desmopuntius johorensis* (Duncker, 1904)** (Fig. 302–39: LBM1210015723, BL 42.0 mm)

One specimen from Thailand (LBM1210015723). Dental formula: 2.3.4–4.3.2; central teeth and minor-row teeth Type 2–2, anterior teeth Type 2–1. Crowns of posterior central and minor-row teeth bent forward with grinding surfaces oriented dorsad. Several denticles present on both margins of grinding surface of all teeth. Crowns of anterior teeth compressed medio-laterally.

***Folifer brevifilis* (Peters, 1881)** (Fig. 302–40: IHCAS210, BL 196.0 mm)

Two specimens from Guangxi (IHCAS210) and Guandong, China (IHCAS802328). Dental formula: 2.3.5–5.3.2; central teeth and minor-row teeth Type 2–2, anterior teeth Type 2–1 or 2–0. Crowns of posterior central and minor-row teeth bent forward with grinding surfaces oriented dorsad. Tooth A2 massive.

***Haludaria fasciatus* (Jerdon, 1849)** (Fig. 302–41: LBM1210013856, BL 62.5 mm)

Three specimens from Thailand (LBM1210013856, 1210013857, and 1210015714). Dental formula: 2.3.4–4(5).3.2; central teeth and minor-row teeth Type 2–2, anterior teeth Type 2–2 or 2–1. Crowns of posterior central and minor-row teeth bent forward with grinding surfaces oriented dorsad. Crowns of Tooth A2 (An2) compressed medio-laterally.

***Hypsibarbus malcolmi* (Smith, 1945)** (Fig. 302–42: LBM1210047490, BL 89.2 mm)

Two specimens from Cambodia (LBM1210047488 and 1210047490). Dental formula: 2.3.5–(4,5).3.2; central teeth and minor-row teeth Type 2–2, anterior teeth Type 2–2 or 2–1. Crowns of posterior central and minor-row teeth bent forward with grinding surfaces oriented dorsad. U-shaped groove on grinding surface of central teeth and minor-row teeth.

***Leptobarbus hoevenii* (Bleeker, 1851)** (Fig. 302–43: LBM1210047610, BL 114.5 mm)

One specimen from Cambodia (LBM1210047610). Dental formula: 2.3.5–5.3.2; central teeth and minor-row teeth Type 2–2, anterior teeth Type 2–0. Crowns of posterior central and minor-row teeth bent forward with grinding surfaces oriented dorsad. Tips of both margins of grinding surfaces of central and minor-row teeth swollen, hump-like.

***Luciocypris langsoni* Vaillant, 1904** (Fig. 302–44: IHCAS635812, BL 214.0 mm)

One specimen from Yunnan, China (IHCAS635812). Dental formula: -4.2.1; central teeth and minor-row teeth Type 2–2, anterior teeth Type 2–2 or 2–1. Crowns of posterior central and minor-row teeth bent forward with grinding surfaces oriented dorsad. Sharp groove present on grinding surface of some teeth.

***Mystacoleucus atridorsalis* Fowler, 1937** (Fig. 302–45: LBM1210047607, BL 60.3 mm)

One specimen from Cambodia (LBM1210047607). Dental formula: 1.3.4–5.3.1; central teeth and minor-row teeth Type 2–2, anterior teeth Type 2–1. Crowns of posterior central and minor-row teeth bent forward with grinding surfaces oriented dorsad.

***Mystacoleucus chilopterus* Fowler, 1935** (Fig. 302–46: LBM1210026925, BL 111.7 mm)

One specimen from Thailand (LBM1210026925). Dental formula: 2.3.4–4.3.2; central teeth and minor-row teeth Type 2–2, anterior teeth Type 2–1. Crowns of posterior central and minor-row teeth bent forward with grinding surfaces oriented dorsad. Shallow U-shaped groove on grinding surface of central teeth and minor-row teeth.

***Mystacoleucus lepturus* Huang, 1979** (Fig. 302–47: IHCAS88111168, BL 77.0 mm)

One specimen from Yunnan, China (IHCAS88111168). Dental formula: 1.2.4–4.2.1; central teeth and minor-row teeth Type 2–2, anterior teeth Type 2–1. Crowns of central teeth and minor-row teeth bent forward with grinding surfaces oriented dorsad. Shallow U-shaped groove on grinding surface of central teeth and minor-row teeth.

***Mystacoleucus marginatus* (Valenciennes, 1842)** (Fig. 302–48: IHCAS92110192, BL 81.0 mm)

One specimen from Yunnan, China (IHCAS92110192). Dental formula: 2.3.4–4.3.1; central teeth and minor-row teeth Type 2–2, anterior teeth Type 2–1. Crowns of central teeth and minor-row teeth bent forward with grinding surfaces oriented dorsad. Shallow U-shaped groove on grinding surface of central teeth and minor-row teeth.

Neolissochilus benasi (**Pellegrin & Chevey, 1936**) (Fig. 302–49: LBM1210037059, BL 345.0 mm)

One specimen from China (LBM1210037059). Dental formula: 2.3.5–5.3.2; central teeth and minor-row teeth Type 2–2, anterior teeth Type 2–2 or 2–1. Crowns of posterior central and minor-row teeth bent forward with grinding surfaces oriented dorsad. U-shaped groove on grinding surface of central teeth and minor-row teeth. Crowns of anterior teeth compressed medio-laterally.

Neolissochilus hexagonolepis (**McClelland, 1839**) (Fig. 302–50: IHCAS7421935, BL 192.0 mm)

Two specimens from Nepal (LBM1210026872) and from Xizang, China (IHCAS7421935). Dental formula: 2.3.5–5.3.2; central teeth and minor-row teeth Type 2–2, anterior teeth Type 2–2 or 2–1. Crowns of central teeth and posterior minor-row teeth bent forward with grinding surfaces oriented dorsad. U-shaped groove on grinding surface of central teeth.

Onychostoma angustistomata (**Fang, 1940**) (Fig. 302–51: IHCAS8760002, BL 107.0 mm)

One specimen from Sichuan, China (IHCAS8760002). Dental formula: 2.3.5–5.3.2; central teeth and minor-row teeth Type 2–2, anterior teeth Type 2–2 or 2–1. Crowns of posterior central and minor-row teeth bent forward with grinding surfaces oriented dorsad. Shallow U-shaped groove on grinding surface of central teeth and minor-row teeth. Both margins of grinding surface of central teeth and minor-row teeth jutting out.

Onychostoma barbatulum (**Pellegrin, 1908**) (Fig. 302–52: LBM1210013968, BL 99.5 mm)

One specimen from Taiwan, China (LBM1210013968). Dental formula: -5.3.2; central teeth and minor-row teeth Type 2–2, anterior teeth Type 2–2 or 2–1. Crowns of posterior central and minor-row teeth bent forward with grinding surfaces oriented dorsad. Shallow U-shaped groove on grinding surface of central teeth and minor-row teeth. Both margins of grinding surface of central teeth and minor-row teeth jutting out.

Onychostoma barbatum (**Lin, 1931**) (Fig. 302–53: IHCAS7493576, BL 170.0 mm)

Two specimens from Zhejiang (IHCAS7493576) and Guangxi, China (IHCAS7542486). Dental formula: 2(1).3.4–(4,5).3.2; central teeth and minor-row teeth Type 2–2, anterior teeth Type 2–2 or 2–1. Crowns of posterior central and minor-row teeth bent forward with grinding surfaces oriented dorsad. Several longitudinal ridges present on grinding surface of central teeth and posterior minor-row teeth. Both margins of grinding surface of central teeth and minor-row teeth jutting out.

Onychostoma elongatum (**Pellegrin & Chevey, 1934**) (Fig. 302–54: IHCAS73102142, BL 138.0 mm)

One specimen from Guangxi, China (IHCAS73102142). Dental formula: -5.3.2; central teeth and minor-row teeth Type 2–2, anterior teeth Type 2–2 or 2–0. Crowns of posterior central and minor-row teeth bent forward with grinding surfaces oriented dorsad. Shallow U-shaped groove on grinding surface of central teeth and minor-row teeth.

Onychostoma gerlachi (**Peters, 1881**) (Fig. 302–55: IHCAS6450078, BL 147.0 mm)

One specimen from Yunnan, China (IHCAS6450078). Dental formula: -4.3.2; central teeth and minor-row teeth Type 2–2, anterior teeth Type 2–1 or 2–0. Crowns of central teeth and minor-row teeth bent forward with grinding surfaces oriented dorsad. Shallow U-shaped groove on grinding surface of central teeth and minor-row teeth.

Onychostoma leptura (**Boulenger, 1900**) (Fig. 302–56: IHCAS7659107, BL 158.0 mm)

One specimen from Guangdong, China (IHCAS7659107). Dental formula: 2.3.5–5.3.2; central teeth and minor-row teeth Type 2–2, anterior teeth Type 2–1 or 2–0. Crowns of central teeth and minor-row teeth bent forward with grinding surfaces oriented dorsad. Shallow U-shaped groove on grinding surface of central teeth and minor-row teeth.

Onychostoma lini (**Wu, 1939**) (Fig. 302–57: IHCAS5877856, BL 177.0 mm)

One specimen from Guangxi, China (IHCAS5877856). Dental formula: 2.3.5–5.3.2; central teeth and minor-row teeth Type 2–2, anterior teeth Type 2–0. Crowns of posterior central and minor-row teeth bent forward with grinding surfaces oriented dorsad. U-shaped groove and longitudinal ridges on grinding surface of central teeth and minor-row teeth. Tooth A2 (An2) massive and relatively large.

***Onychostoma macrolepis* (Bleeker, 1871)** (Fig. 302-58: IHCAS8252311, BL 151.0 mm)

One specimen from Gansu, China (IHCAS8252311). Dental formula: 2.3.5-5.3.2; central teeth and minor-row teeth Type 2-2, anterior teeth Type 2-2 or 2-1. Crowns of posterior central and minor-row teeth bent forward with grinding surfaces oriented dorsad. Some teeth well worn.

***Onychostoma ovale* Pellegrin & Chevey, 1936** (Fig. 302-59: IHCAS655030, BL 276.0 mm)

One specimen from Sichuan, China (IHCAS655030). Dental formula: 2.3.5-5.3.1; central teeth and minor-row teeth Type 2-2, anterior teeth Type 2-2 or 2-1. Crowns of central teeth and minor-row teeth bent forward with grinding surfaces oriented dorsad. U-shaped groove on grinding surface of central teeth and minor-row teeth.

***Onychostoma rarum* (Lin, 1933)** (Fig. 302-60: IHCAS3745, BL 214.0 mm)

One specimen from Hunan, China (IHCAS3745). Dental formula: 2.3.5-5.3.2; central teeth and minor-row teeth Type 2-2, anterior teeth Type 2-2 or 2-1. Crowns of posterior central and minor-row teeth bent forward with grinding surfaces oriented dorsad. U-shaped groove and longitudinal ridges on grinding surface of central teeth and minor-row teeth.

***Onychostoma simum* (Sauvage & Dabry de Thiersant, 1874)** (Fig. 302-61: IHCAS6640550, BL 127.0 mm)

One specimen from Guizhou, China (IHCAS6640550). Dental formula: 2.3.5-5.3.2; central teeth and minor-row teeth Type 2-2, anterior teeth Type 2-2 or 2-1. Crowns of central teeth and minor-row teeth bent forward with grinding surfaces oriented dorsad. U-shaped groove on grinding surface of central teeth and minor-row teeth.

***Oreichthys cosuatis* (Hamilton, 1822)** (Fig. 302-62: LBM1210015693, BL 31.6 mm)

One specimen from India (LBM1210015693). Dental formula: 2.3.5-4.3.2; central teeth and minor-row teeth Type 2-2, anterior teeth Type 2-1 or 2-0. Crowns of central teeth and minor-row teeth bent forward with grinding surfaces oriented dorsad.

***Osteobrama cotio* (Hamilton, 1822)** (Fig. 302-63: LBM1210014002, BL 71.1 mm)

One specimen from Nepal (LBM1210014002).

Dental formula: 2.3.4-4.3.2; central teeth and minor-row teeth Type 2-2, anterior teeth Type 2-1 or 2-0. Crowns of posterior central and minor-row teeth bent forward with grinding surfaces oriented dorsad. Shallow U-shaped groove on grinding surface of central teeth and minor-row teeth.

***Poropuntius chonglingchungi* (Tchang, 1938)** (Fig. 302-64: IHCAS8960013, BL 254.0 mm)

One specimen from Yunnan, China (IHCAS8960013). Dental formula: 2.3.5-5.3.2; central teeth and minor-row teeth Type 2-2, anterior teeth Type 2-0. Crowns of posterior central teeth and minor-row teeth bent forward with grinding surfaces oriented dorsad. Shallow U-shaped groove on grinding surface of central teeth and minor-row teeth. Tooth A2 massive.

***Poropuntius daliensis* (Wu & Lin, 1977)** (Fig. 302-65: IHCAS646935, BL 208.0 mm)

One specimen from Yunnan, China (IHCAS646935). Dental formula: 2.3.5-5.3.2; central teeth and minor-row teeth Type 2-2, anterior teeth Type 2-1. Crowns of central teeth and posterior minor-row teeth bent forward with grinding surfaces oriented dorsad. Broad and shallow U-shaped groove on grinding surface of central teeth and minor-row teeth.

***Poropuntius deauratus* (Valenciennes, 1842)** (Fig. 302-66: LBM1210047426, BL 113.6 mm)

One specimen from Cambodia (LBM1210047426). Dental formula: -5.3.2; central teeth and minor-row teeth Type 2-2, anterior teeth Type 2-2 or 2-1. Crowns of posterior central and minor-row teeth bent forward with grinding surfaces oriented dorsad. Shallow U-shaped groove on grinding surface of central teeth and minor-row teeth.

***Poropuntius huangchuchieni* (Tchang, 1962)** (Fig. 302-67: IHCAS6530986, BL 158.0 mm)

One specimen from Yunnan, China (IHCAS6530986). Dental formula: 2.3.5-5.3.2; central teeth and minor-row teeth Type 2-2, anterior teeth Type 2-1 or 2-0. Crowns of posterior central and minor-row teeth bent forward with grinding surfaces oriented dorsad. Shallow U-shaped groove on grinding surface of central teeth and minor-row teeth.

***Poropuntius ikedai* (Harada, 1943)** (Fig. 302‒68: IHCAS7655087, BL 95.0 mm)

One specimen from Hainan, China (IHCAS7655087). Dental formula: 2.3.5‒5.3.2; central teeth and minor-row teeth Type 2‒2, anterior teeth Type 2‒1 or 2‒0. Crowns of posterior central and minor-row teeth bent forward with grinding surfaces oriented dorsad. Shallow U-shaped groove on grinding surface of central teeth and minor-row teeth.

***Poropuntius krempfi* (Pellegrin & Chevey, 1934)** (Fig. 302‒69: IHCAS6440552, BL 142.0 mm)

One specimen from Yunnan, China (IHCAS6440552). Dental formula: 2.3.5‒4.3.2; central teeth and minor-row teeth Type 2‒2, anterior teeth Type 2‒1 or 2‒0. Crowns of posterior central and minor-row teeth bent forward with grinding surfaces oriented dorsad. Shallow U-shaped groove on grinding surface of central teeth and minor-row teeth. Some teeth well worn.

***Poropuntius opisthopterus* (Wu, 1977)** (Fig. 302‒70: IHCAS904437, BL 121.0 mm)

One specimen from Yunnan, China (IHCAS904437). Dental formula: 2.3.5‒5.3.2; central teeth and minor-row teeth Type 2‒2, anterior teeth Type 2‒1 or 2‒0. Crowns of posterior central and minor-row teeth bent forward with grinding surfaces oriented dorsad. Shallow U-shaped groove on grinding surface of central teeth and minor-row teeth.

***Sinocyclocheilus grahami* (Regan, 1904)** (Fig. 302‒71: IHCAS634369, BL 117.0 mm)

One specimen from Yunnan, China (IHCAS634369). Dental formula: 2.3.4‒4.3.2; central teeth and minor-row teeth Type 2‒2, anterior teeth Type 2‒1. Crowns of posterior central and minor-row teeth bent forward with grinding surfaces oriented dorsad. Fine longitudinal ridges present on grinding surface of posterior central teeth and minor-row teeth.

***Sinocyclocheilus tingi* Fang, 1936** (Fig. 302‒72: IHCAS636482, BL 193.0 mm)

One specimen from Yunnan, China (IHCAS636482). Dental formula: 2.3.4‒4.3.2; central teeth and minor-row teeth Type 2‒2, anterior teeth Type 2‒2 or 2‒1. Crowns of posterior central and minor-row teeth bent forward with grinding surfaces oriented dorsad. Several fine parallel longitudinal ridges present on grinding surface of central

teeth and minor-row teeth.

***Sinocyclocheilus yangzongensis* Tsü & Chen, 1977** (Fig. 302‒73: IHCAS6351014, BL 115.0 mm)

One specimen from Yunnan, China (IHCAS6351014). Dental formula: 2.3.4‒4.3.1; central teeth and minor-row teeth Type 2‒2, anterior teeth Type 2‒2 or 2‒1. Crowns of posterior central and minor-row teeth bent forward with grinding surfaces oriented dorsad. Longitudinal ridges present on grinding surface of central teeth and minor-row teeth.

***Spinibarbus denticulatus* (Oshima, 1926)** (Fig. 302‒74: IHCAS587821, BL 267.0 mm)

Two specimens from Guangxi (IHCAS587821) and Guandong, China (IHCAS602176). Dental formula: 2.3.5‒5.3.2; central teeth and minor-row teeth Type 2‒2, anterior teeth Type 2‒1 or 2‒0. Crowns of posterior central and minor-row teeth bent forward with grinding surfaces oriented dorsad. U-shaped groove on grinding surface of central teeth and minor-row teeth.

***Spinibarbus holandi* Oshima, 1919** (Fig. 302‒75: IHCAS7640004, BL 213.0 mm)

One specimen from Guangdong, China (IHCAS7640004). Dental formula: 2.3.5‒5.3.2; central teeth and minor-row teeth Type 2‒2, anterior teeth Type 2‒1 or 2‒0. Crowns of posterior central and minor-row teeth bent forward with grinding surfaces oriented dorsad. U-shaped groove on grinding surface of central teeth and minor-row teeth.

***Spinibarbus sinensis* (Bleeker, 1871)** (Fig. 302‒76: IHCAS6650260, BL 164.0 mm)

Nine specimens from Guizhou, China (IHCAS6650260, LBM1210016154, 1210016155, 1210016157, 1210020551,1210020552, 1210021274, 1210021277, and 1210027979). Dental formula: 2(1).3.5(4)-5(4).3.2; central teeth and minor-row teeth Type 2‒2, anterior teeth Type 2‒1 or 2‒0. Crowns of posterior central and minor-row teeth bent forward with grinding surfaces oriented dorsad. U-shaped groove on grinding surface of central teeth and minor-row teeth.

***Systomus orphoides* (Valenciennes, 1842)** (Fig. 302‒77: LBM1210014047, BL 88.0 mm)

One specimen from Thailand (LBM1210014047). Dental formula: -4.3.2; central teeth and minor-row teeth

Type 2-2, anterior teeth Type 2-1 or 2-0. Crowns of posterior central and minor-row teeth bent forward with grinding surfaces oriented dorsad. U-shaped groove on grinding surface of central teeth and minor-row teeth. Hook and tips of both margins of central teeth swollen, hump-like.

Tor douronensis (**Valenciennes, 1842**) (Fig. 302-78: IHCAS7840017, BL 172.0 mm)

One specimen from Yunnan, China (IHCAS7840017). Dental formula: 2.3.5-5.3.2; central teeth and minor-row teeth Type 2-2, anterior teeth Type 2-1 or 2-0. Crowns of posterior central and minor-row teeth bent forward with grinding surfaces oriented dorsad. U-shaped groove on grinding surface of central teeth and minor-row teeth. Tooth crowns well worn.

Tor putitora (**Hamilton, 1822**) (Fig. 302-79: LBM1210047612, BL 47.9 mm)

One specimen from Nepal (LBM1210047612). Dental formula: -5.3.2; central teeth and minor-row teeth Type 2-2, anterior teeth Type 2-2 or 2-1. Crowns of posterior central and minor-row teeth bent forward with grinding surfaces oriented dorsad. U-shaped groove and marginal ridge on grinding surface of central teeth. Tooth A2 (An2) compresed medio-laterally.

Tor sinensis **Wu, 1977** (Fig. 302-80: IHCAS92110037, BL 133.0 mm)

One specimen from Yunnan, China (IHCAS92110037). Dental formula: 2.3.5-5.3.2; central teeth and minor-row teeth Type 2-2, anterior teeth Type 2-0. Crowns of posterior central and minor-row teeth bent forward with grinding surfaces oriented dorsad. U-shaped groove on grinding surface of central teeth and minor-row teeth. Some teeth well worn.

Tor soro (**Valenciennes, 1842**) (Fig. 302-81: LBM1210014055, BL 85.0 mm)

One specimen from Thailand (LBM1210014055). Dental formula: 2.3.5-5.3.2; central teeth and minor-row teeth Type 2-2, anterior teeth Type 2-2 or 2-1. Crowns of posterior central and minor-row teeth bent forward with grinding surfaces oriented dorsad. U-shaped groove on grinding surface of central teeth and minor-row teeth.

Puntius arulius (**Jerdon, 1849**) (Fig. 302-82: LBM1210015716, BL 45.3 mm)

One specimen from Thailand (LBM1210015716). Dental formula: 2.3.5-5.3.2; central teeth and minor-row teeth Type 2-2, anterior teeth Type 2-1 or 2-0. Crowns of central teeth and minor-row teeth bent forward with grinding surfaces oriented dorsad.

Puntius aurotaeniatus (**Tirant, 1885**) (Fig. 302-83: LBM1210026902, BL 50.1 mm)

Two specimens from Cambodia (LBM1210026902 and 1210026922). Dental formula: (1,2).3.(4,5)-5.3.2; central teeth and minor-row teeth Type 2-2, anterior teeth Type 2-1 or 2-0. Crowns of central teeth and posterior minor-row teeth bent forward with grinding surfaces oriented dorsad. Shallow U-shaped groove on grinding surface of central teeth and minor-row teeth. Crowns of tooth A2 compressed medio-laterally.

Puntius binotatus (**Valenciennes, 1842**) (Fig. 302-84: LBM1210014051, BL 19.4 mm)

Two specimens from Thailand (LBM1210014050 and 1210014051). Dental formula: 1.3.5-5.3.(1,2); central teeth and minor-row teeth Type 2-2, anterior teeth Type 2-1 or 2-0. Crowns of posterior central and minor-row teeth bent forward with grinding surfaces oriented dorsad. Both margins of grinding surface of central teeth and posterior minor-row teeth jutting out.

Puntius conchonius (**Hamilton, 1822**) (Fig. 302-85: LBM1210015699, BL 38.3 mm)

One specimen from Nepal (LBM1210014004 and 1210015699). Dental formula: 2.3.5-5.3.2; central teeth and minor-row teeth Type 2-2, anterior teeth Type 2-0. Crowns of central teeth and minor-row teeth bent forward with grinding surfaces oriented dorsad. Both margins of grinding surface of central teeth and posterior minor-row teeth jutting out.

Puntius lateristriga (**Valenciennes, 1842**) (Fig. 302-86: LBM1210024515, BL 62.6 mm)

Two specimens from Thailand (LBM1210024513 and 1210024515). Dental formula: 2.3.4-4.3.2; central teeth and minor-row teeth Type 2-2, anterior teeth Type 2-1. Crowns of posterior central and minor-row teeth bent forward with grinding surfaces oriented dorsad.

Puntius oligolepis **(Bleeker, 1853)** (Fig. 302-87: LBM1210015743, BL 28.0 mm)

Two specimens from Indonesia (LBM1210015709 and 1210015743). Dental formula: 2.3.5-(4,5).3.2; central teeth and minor-row teeth Type 2-2, anterior teeth Type 2-1 or 2-0. Crowns of posterior central and minor-row teeth bent forward with grinding surfaces oriented dorsad. Both margins jutting out and ridge present on grinding surface of posterior central and minor-row teeth.

Puntius partipentazona **(Fowler, 1934)** (Fig. 302-88: LBM1210024417, BL 31.5 mm)

One specimen from Malaysia (LBM1210024417). Dental formula: 2.3.4-4.3.2; central teeth and minor-row teeth Type 2-2, anterior teeth Type 2-1. Crowns of posterior central and minor-row teeth bent forward with grinding surfaces oriented dorsad.

Puntius rhomboocellatus **Koumans, 1940** (Fig. 302-89: LBM1210015722, BL 31.7 mm)

One specimen from Indonesia (LBM1210015722). Dental formula: 2.3.4-4.3.2; central teeth and minor-row teeth Type 2-2, anterior teeth Type 2-1. Crowns of posterior central and minor-row teeth bent forward with grinding surfaces oriented dorsad. Several denticles present on both margins of grinding surface of central teeth and minor-row teeth. Tooth A1 (An2) compressed medio-laterally.

Puntius sealei **(Herre, 1933)** (Fig. 302-90: LBM1210026876, BL 65.1 mm)

One specimen from Cambodia (LBM1210026876). Dental formula: -5.3.2; central teeth and minor-row teeth Type 2-2, anterior teeth Type 2-1 or 2-0. Crowns of posterior central and minor-row teeth bent forward with grinding surfaces oriented dorsad. Tooth A2 (An2) compressed medio-laterally.

Puntius tetrazona **(Bleeker, 1855)** (Fig. 302-91: LBM1210015705, BL 35.5 mm)

Two specimens from Indonesia (LBM1210015705 and 1210015742). Dental formula: 2.3.5-4.3.2; central teeth and minor-row teeth Type 2-2, anterior teeth Type 2-1 or 2-0. Crowns of posterior central and minor-row teeth bent forward with grinding surfaces oriented dorsad. Longitudinal ridge present on grinding surface of central teeth and minor-row teeth.

Puntius nigrofasciatus **(Günther, 1868)** (Fig. 302-92: LBM1210015707, BL 35.0 mm)

One specimen from Sri Lanka (LBM1210015707). Dental formula: 1.3.4-4.3.2; central teeth and minor-row teeth Type 2-2, anterior teeth Type 2-1. Crowns of central teeth and minor-row teeth bent forward with grinding surfaces oriented dorsad. Both margins jutting out and several fine parallel ridges present on grinding surface of central teeth and minor-row teeth.

Puntius titteya **Deraniyagala, 1929** (Fig. 302-93: LBM1210015725, BL 23.9 mm)

Two specimens from Sri Lanka (LBM1210015725 and 1210024420). Dental formula: 1.3.(4,5)-4.3.2; central teeth and minor-row teeth Type 2-2, anterior teeth Type 2-1 or 2-0. Crowns of posterior central and minor-row teeth bent forward with grinding surfaces oriented dorsad. Tooth A1 (An2) or A2 (An2) compressed medio-laterally.

Puntius gelius **(Hamilton, 1822)** (Fig. 302-94: LBM1210015718, BL 23.2 mm)

One specimen from India (LBM1210015718). Dental formula: 2.3.4-4.3.2; all teeth Type 2-2. Crowns of posterior central and minor-row teeth bent forward with grinding surfaces oriented dorsad.

Puntius sophore **(Hamilton, 1822)** (Fig. 302-95: LBM1210014003, BL 50.9 mm)

Three specimens from Nepal (LBM1210013994, 1210014003, and 1210014794). Dental formula: 2.3.5-5.3.2; central teeth and minor-row teeth Type 2-2, anterior teeth Type 2-2 or 2-0. Crowns of posterior central and minor-row teeth bent forward with grinding surfaces oriented dorsad. Posterior minor-row teeth compressed antero-posteriorly. Anterior minor-row teeth and tooth A2 (An2) compressed medio-laterally. Some teeth worn.

Puntius chola **(Hamilton, 1822)** (Fig. 302-96: LBM1210013998, BL 82.0 mm)

One specimen from Nepal (LBM1210013998). Dental formula: - 5.3.1; central teeth and minor-row teeth Type 2-2, anterior teeth Type 2-1 or 2-0. Crowns of posterior central and minor-row teeth bent forward with grinding surfaces oriented dorsad. Posterior minor-row teeth compressed antero-posteriorly. Anterior minor-row teeth and tooth A2 (An2) compressed medio-laterally. Some teeth worn.

Puntius brevis (**Bleeker, 1849**) (Fig. 302‑97: LBM1210026908, BL 62.5 mm)

One specimen from Cambodia (LBM1210026908). Dental formula: 2.3.5‑5.3.2; central teeth and minor-row teeth Type 2‑2, anterior teeth Type 2‑2 or 2‑0. Crowns of posterior central and minor-row teeth bent forward with grinding surfaces oriented dorsad. Transverse groove on grinding surface of central teeth and minor-row teeth. Posterior central and minor-row teeth compressed antero-posteriorly. Some teeth worn.

Puntius bimaculatus (**Bleeker, 1863**) (Fig. 302‑98: LBM1210015713, BL 46.5 mm)

One specimen from Sri Lanka (LBM1210015713). Dental formula: 1.3.5‑4.2.2; central teeth and minor-row teeth Type 2‑2, anterior teeth Type 2‑0. Crowns of posterior central and minor-row teeth bent forward with grinding surfaces oriented dorsad. Transverse groove on grinding surface of central teeth and minor-row teeth. Posterior central and minor-row teeth compressed antero-posteriorly. Some teeth worn.

Puntius vittatus **Day, 1865** (Fig. 302‑99: LBM1210015701, BL 21.5 mm)

One specimen from India (LBM1210015701). Dental formula: 1.3.4‑4.3.1; central teeth and minor-row teeth Type 2‑2, anterior teeth Type 2‑0. Crowns of posterior central and minor-row teeth bent forward with grinding surfaces oriented dorsad. Transverse groove on grinding surface of central teeth and minor-row teeth. Posterior central and minor-row teeth compressed antero-posteriorly. Some teeth worn.

Parator zonatus (**Lin, 1935**) (Fig. 302‑100: IHCAS587185, BL 137.0 mm)

One specimen from Guangxi, China (IHCAS587185). Dental formula: 2.3.5‑5.3.2; central teeth and minor-row teeth Type 2‑2, anterior teeth Type 2‑1 or 2‑0. Crowns of posterior central and minor-row teeth bent forward with grinding surfaces oriented dorsad. Transverse or U-shaped groove on grinding surface of central teeth and minor-row teeth. Some teeth worn.

Capoeta capoeta gracilis (**Keyserling, 1861**) (Fig. 302‑101: LBM1210013969, BL 117.0 mm)

One specimen from Iran (LBM1210013969). Dental formula: 2.3.5‑4.3.1; central teeth and minor-row teeth Type 2‑2, anterior teeth Type 2‑2 or 2‑1. Crowns of central teeth and minor-row teeth bent forward with grinding surfaces oriented dorsad and compressed antero posteriorly. Crowns well worn with secondary grinding surfaces.

Catlocarpio siamensis **Boulenger, 1898** (Fig. 302‑102: LBM1210047433, BL 500.0 mm)

Two specimens from Laos (LBM1210047433) and Cambodia (LBM1210047424). Dental formula: 4‑4; central teeth Type 4‑1 or 5‑0, anterior teeth Type 2‑1. Crowns of posterior central teeth bent forward. Tiny hamulus groove present on medial side of crown of central teeth. Many denticles present on grinding surface of central teeth.

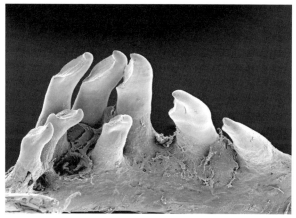

Fig. 302-1. *Hampala dispar*: LBM1210013989.

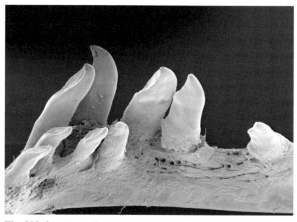

Fig. 302-2. *Hampala macrolepidota*: LBM1210013990.

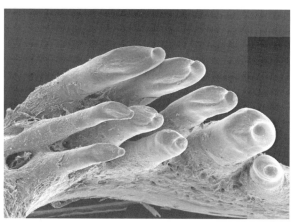

Fig. 302-3. *Percocypris pingi*: IHCAS8180235.

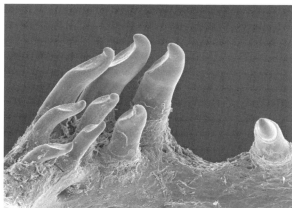

Fig. 302-4. *Percocypris regani*: IHCAS81104166.

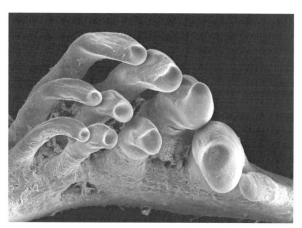

Fig. 302-5. *Acrossocheilus beijiangensis*: IHCAS874480.

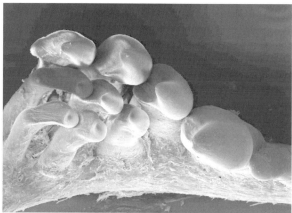

Fig. 302-6. *Acrossocheilus fasciatus*: IHCAS59122.

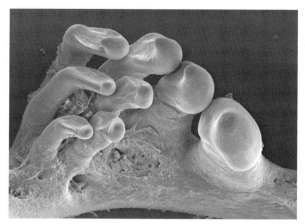

Fig. 302-7. *Acrossocheilus iridescens*: IHCAS602243.

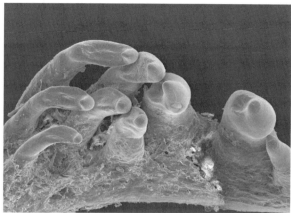

Fig. 302-8. *Acrossocheilus kreyenbergii*: IHCAS839232.

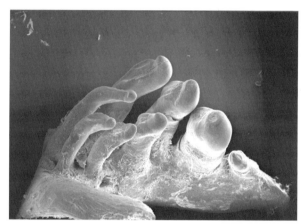

Fig. 302-9. *Acrossocheilus longipinnis*: IHCAS875927.

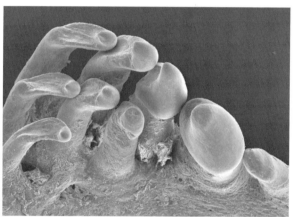

Fig. 302-10. *Acrossocheilus wenchowensis*: IHCAS7461127.

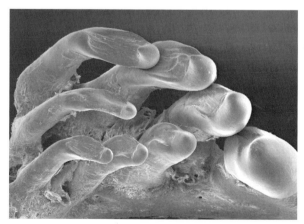

Fig. 302-11. *Acrossocheilus hemispinus*: IHCAS81104633.

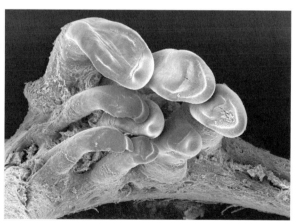

Fig. 302-12. *Acrossocheilus monticola*: IHCAS8061243.

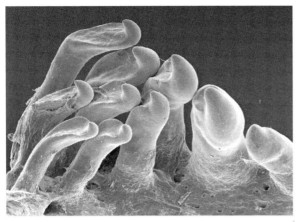

Fig. 302-13. *Acrossocheilus paradoxus*: LBM1210024283.

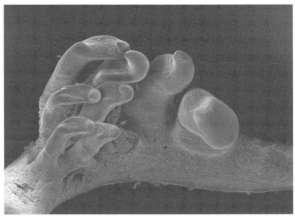

Fig. 302-14. *Acrossocheilus parallens*: IHCAS9051943.

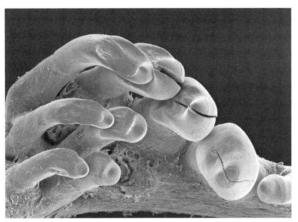

Fig. 302-15. *Acrossocheilus yunnanensis*: IHCAS8180252.

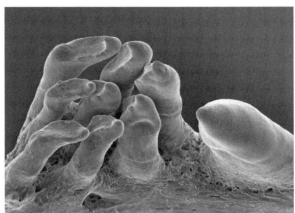

Fig. 302-16. *Anematichthys armatus*: LBM1210026904.

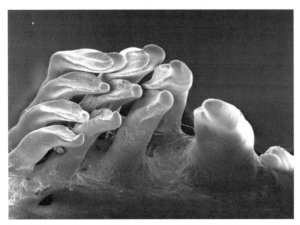

Fig. 302-17. *Anematichthys repasson*: LBM1210013849.

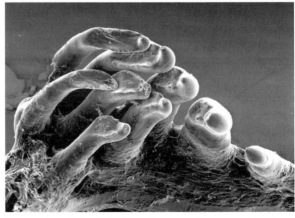

Fig. 302-18. *Balantiocheilos melanopterus*: LBM1210014973.

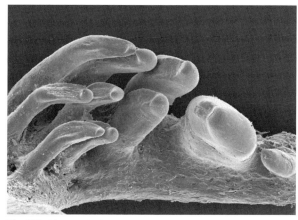

Fig. 302-19. *Barbodes semifasciolatus*: IHCAS8243308.

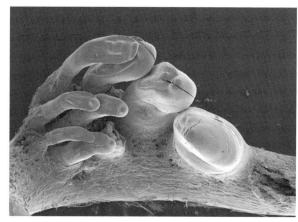

Fig. 302-20. *Barbodes wynaadensis*: IHCAS58530.

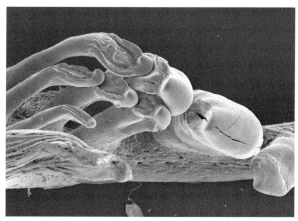

Fig. 302-21. *Barbonymus altus*: LBM1210026933.

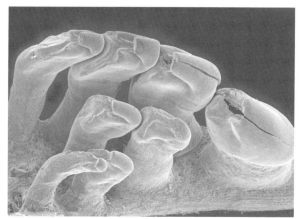

Fig. 302-22. *Barbonymus gonionotus*: LBM1210047432.

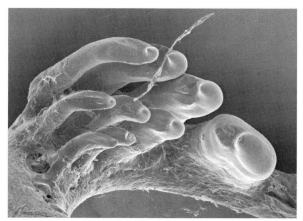

Fig. 302-23. *Barbus bifrenatus*: LBM1210024287.

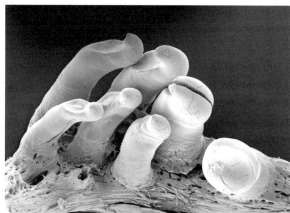

Fig. 302-24. *Barbus tauricus*: LBM1210013972.

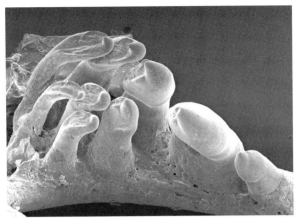

Fig. 302−25. *Barbus afrohamiltoni*: LBM1210024285.

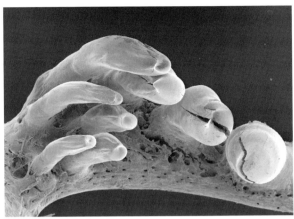

Fig. 302−26. *Barbus barbus*: LBM1210013971.

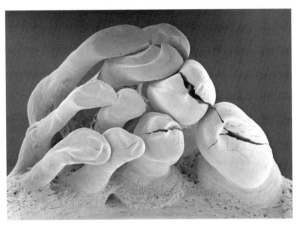

Fig. 302−27. *Barbus bynni*: LBM1210014045.

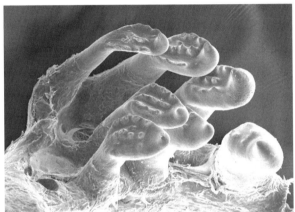

Fig. 302−28. *Barbus jae*: LBM1210047615.

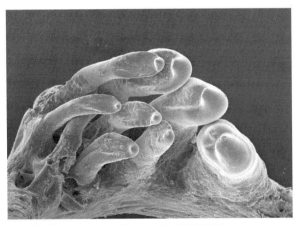

Fig. 302−29. *Barbus paludinosus*: LBM1210024291.

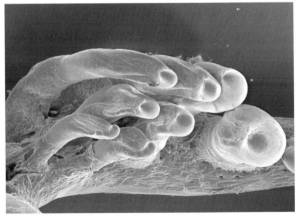

Fig. 302−30. *Barbus peloponnesius*: LBM1210013842.

Fig. 302–31. *Barbus toppini*: LBM1210024293.

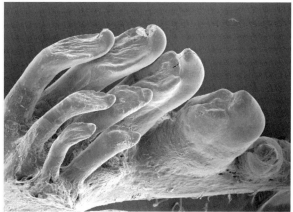

Fig. 302–32. *Barbus trimaculatus*: LBM1210024294.

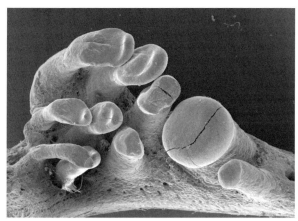

Fig. 302–33. *Cosmochilus harmandi*: LBM1210047608.

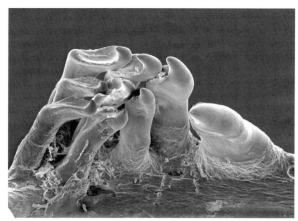

Fig. 302–34. *Cyclocheilichthys apogon*: LBM1210013848.

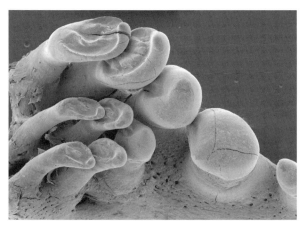

Fig. 302–35. *Cyclocheilichthys enoplos*: LBM1210014064.

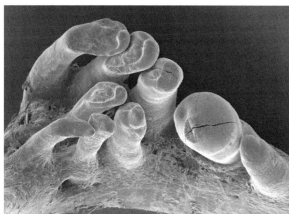

Fig. 302–36. *Cyclocheilichthys furcatus*: LBM1210047486.

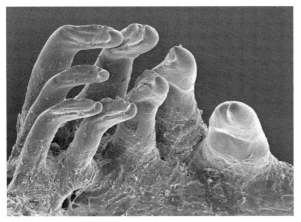

Fig. 302-37. *Cyclocheilichthys lagleri*: LBM1210026943.

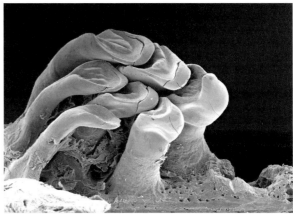

Fig. 302-38. *Cyprinion kais*: LBM1210013984.

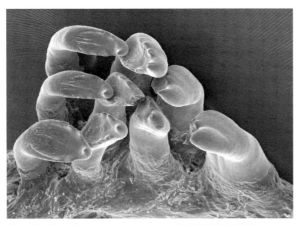

Fig. 302-39. *Desmopuntius johorensis*: LBM1210015723.

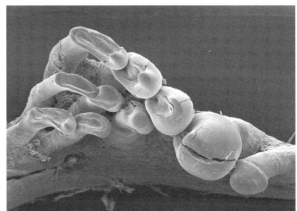

Fig. 302-40. *Folifer brevifilis*: IHCAS210.

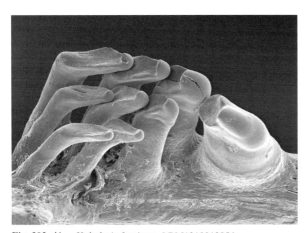

Fig. 302-41. *Haludaria fasciatus*: LBM1210013856.

Fig. 302-42. *Hypsibarbus malcolmi*: LBM1210047490.

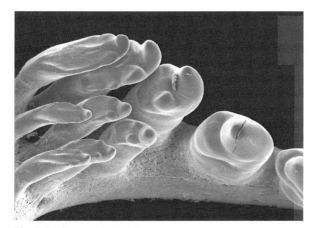

Fig. 302−43. *Leptobarbus hoevenii*: LBM1210047610.

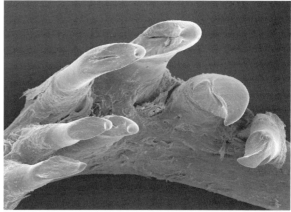

Fig. 302−44. *Luciocypris langsoni*: IHCAS635812.

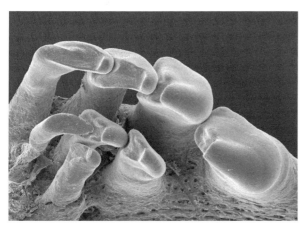

Fig. 302−45. *Mystacoleucus atridorsalis*: LBM1210047607.

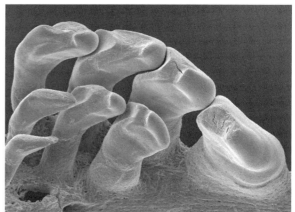

Fig. 302−46. *Mystacoleucus chilopterus*: LBM1210026925.

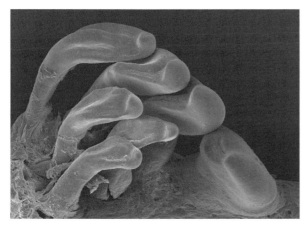

Fig. 302−47. *Mystacoleucus lepturus*: IHCAS88111168.

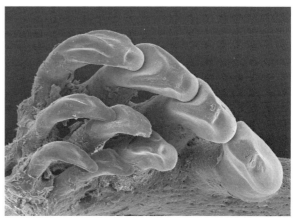

Fig. 302−48. *Mystacoleucus marginatus*: IHCAS92110192.

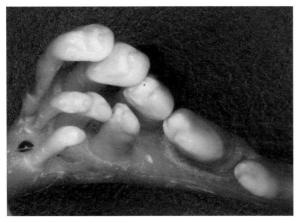

Fig. 302-49. *Neolissochilus benasi*: LBM1210037059.

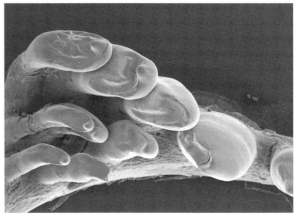

Fig. 302-50. *Neolissochilus hexagonolepis*: IHCAS7421935.

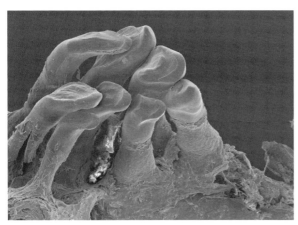

Fig. 302-51. *Onychostoma angustistomata*: IHCAS8760002.

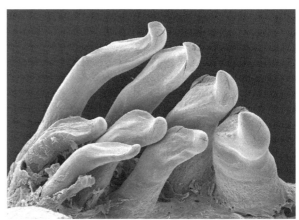

Fig. 302-52. *Onychostoma barbatulum*: LBM1210013968.

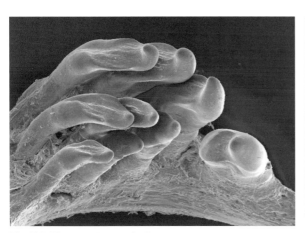

Fig. 302-53. *Onychostoma barbatum*: IHCAS7493576.

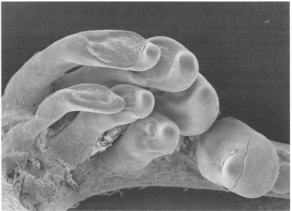

Fig. 302-54. *Onychostoma elongatum*: IHCAS73102142.

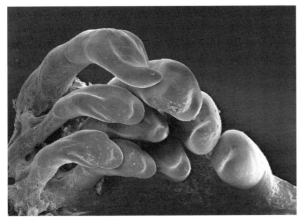

Fig. 302-55. *Onychostoma gerlachi*: IHCAS6450078.

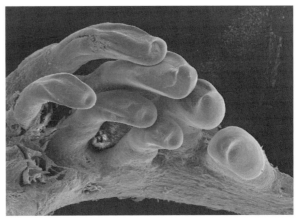

Fig. 302-56. *Onychostoma leptura*: IHCAS7659107.

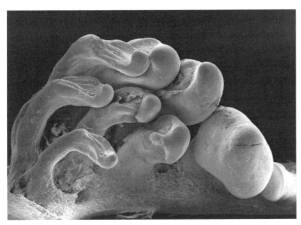

Fig. 302-57. *Onychostoma lini*: IHCAS5877856.

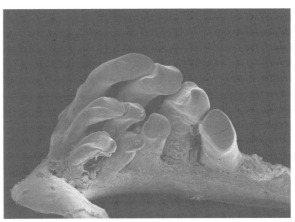

Fig. 302-58. *Onychostoma macrolepis*: IHCAS8252311.

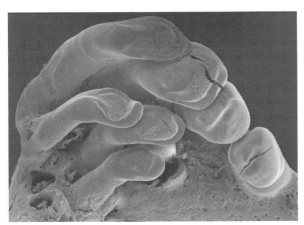

Fig. 302-59. *Onychostoma ovale*: IHCAS655030.

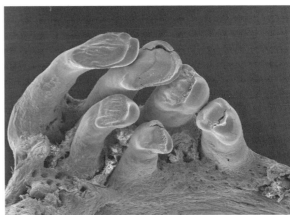

Fig. 302-60. *Onychostoma rarum*: IHCAS3745.

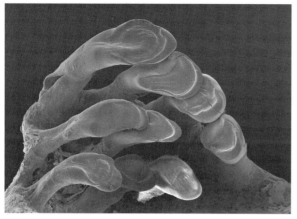

Fig. 302-61. *Onychostoma simum*: IHCAS6640550.

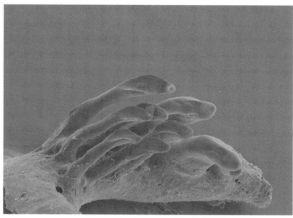

Fig. 302-62. *Oreichthys cosuatis*: LBM1210015693.

Fig. 302-63. *Osteobrama cotio*: LBM1210014002.

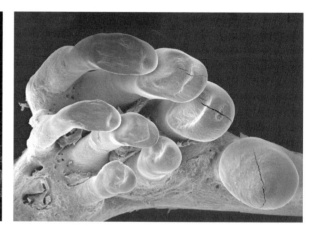

Fig. 302-64. *Poropuntius chonglingchungi*: IHCAS8960013.

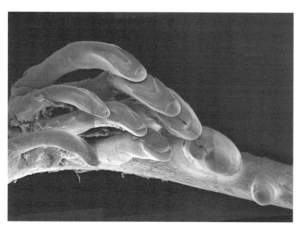

Fig. 302-65. *Poropuntius daliensis*: IHCAS646935.

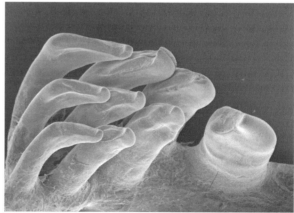

Fig. 302-66. *Poropuntius deauratus*: LBM1210047426.

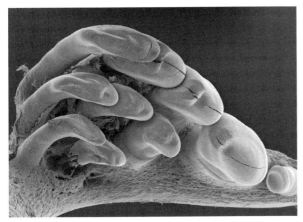

Fig. 302-67. *Poropuntius huangchuchieni*: IHCAS6530986.

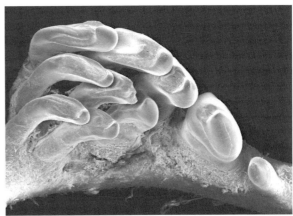

Fig. 302-68. *Poropuntius ikedai*: IHCAS7655087.

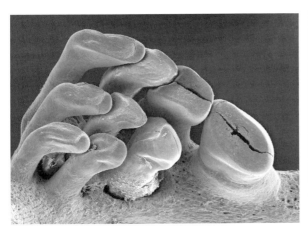

Fig. 302-69. *Poropuntius krempfi*: IHCAS6440552.

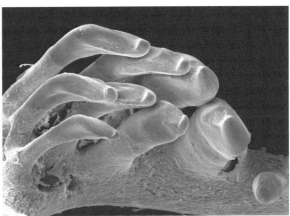

Fig. 302-70. *Poropuntius opisthopterus*: IHCAS904437.

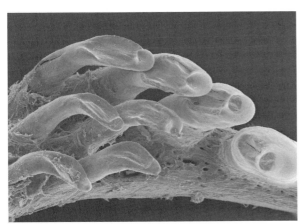

Fig. 302-71. *Sinocyclocheilus grahami*: IHCAS634369.

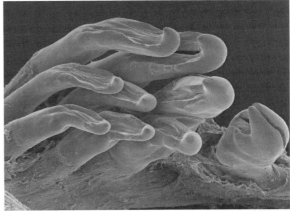

Fig. 302-72. *Sinocyclocheilus tingi*: IHCAS636482.

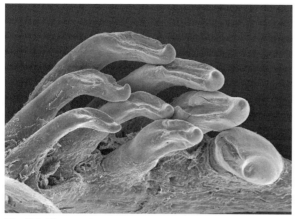

Fig. 302‾73. *Sinocyclocheilus yangzongensis*: IHCAS6351014.

Fig. 302‾74. *Spinibarbus denticulatus*: IHCAS587821.

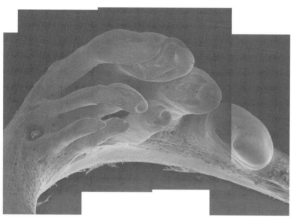

Fig. 302‾75. *Spinibarbus holandi*: IHCAS7640004.

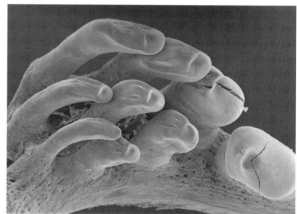

Fig. 302‾76. *Spinibarbus sinensis*: IHCAS6650260.

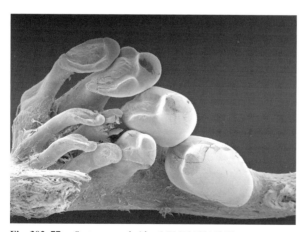

Fig. 302‾77. *Systomus orphoides*: LBM1210014047.

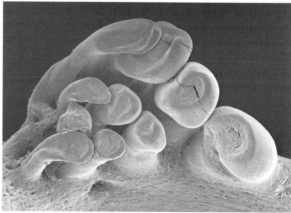

Fig. 302‾78. *Tor douronensis*: IHCAS7840017.

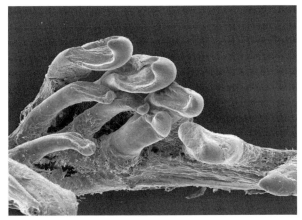

Fig. 302-79. *Tor putitora*: LBM1210047612.

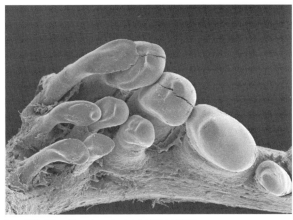

Fig. 302-80. *Tor sinensis*: IHCAS92110037.

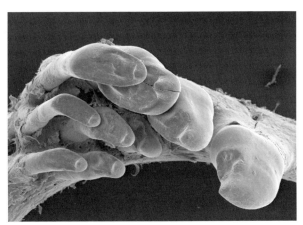

Fig. 302-81. *Tor soro*: LBM1210014055.

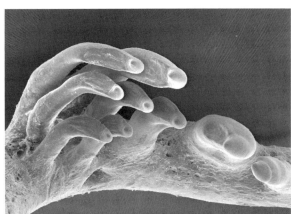

Fig. 302-82. *Puntius arulius*: LBM1210015716.

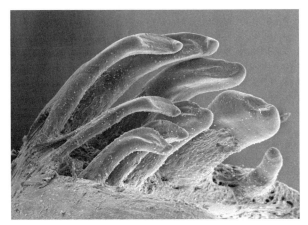

Fig. 302-83. *Puntius aurotaeniatus*: LBM1210026902.

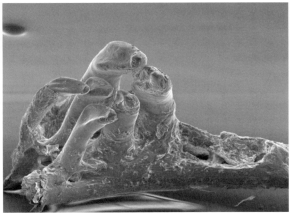

Fig. 302-84. *Puntius binotatus*: LBM1210014051.

3. Descriptions of the pharyngeal dentition of cyprinid subfamilies 69

Fig. 302-85. *Puntius conchonius*: LBM1210015699.

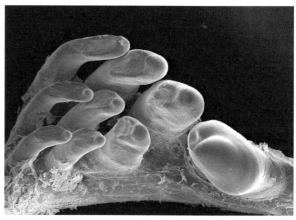

Fig. 302-86. *Puntius lateristriga*: LBM1210024515.

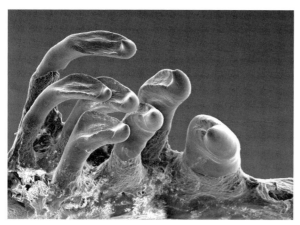

Fig. 302-87. *Puntius oligolepis*: LBM1210015743.

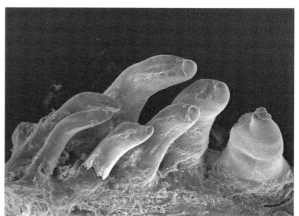

Fig. 302-88. *Puntius partipentazona*: LBM1210024417.

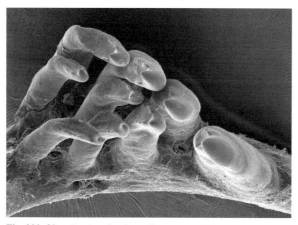

Fig. 302-89. *Puntius rhomboocellatus*: LBM1210015722.

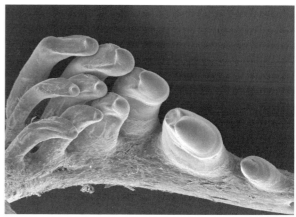

Fig. 302-90. *Puntius sealei*: LBM1210026876.

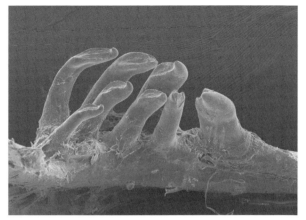

Fig. 302-91. *Puntius tetrazona*: LBM1210015705.

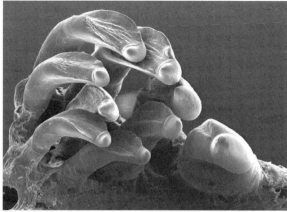

Fig. 302-92. *Puntius nigrofasciatus*: LBM1210015707.

Fig. 302-93. *Puntius titteya*: LBM1210015725.

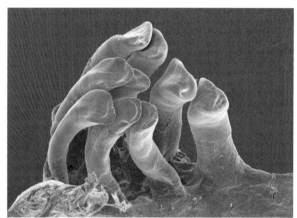

Fig. 302-94. *Puntius gelius*: LBM1210015718.

Fig. 302-95. *Puntius sophore*: LBM1210014003.

Fig. 302-96. *Puntius chola*: LBM1210013998.

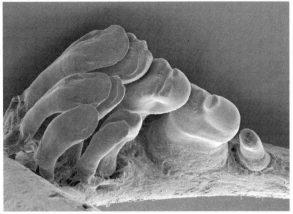

Fig. 302-97. *Puntius brevis*: LBM1210026908.

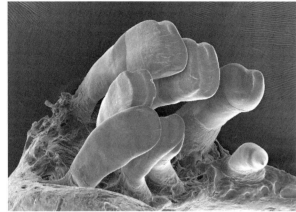

Fig. 302-98. *Puntius bimaculatus*: LBM1210015713.

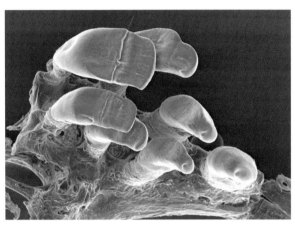

Fig. 302-99. *Puntius vittatus*: LBM1210015701.

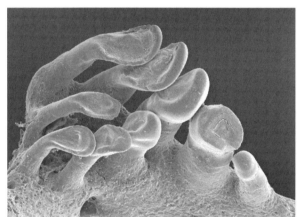

Fig. 302-100. *Parator zonatus*: IHCAS587185.

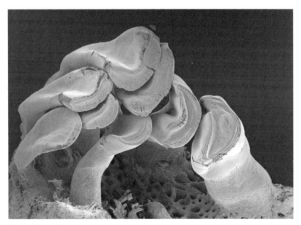

Fig. 302-101. *Capoeta capoeta gracilis*: LBM1210013969.

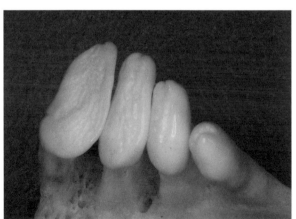

Fig. 302-102. *Catlocarpio siamensis*: LBM1210047433.

3-3. Schizothoracinae

The pharyngeal dentitions of 20 species in five genera of Schizothoracinae were examined. Their teeth are arranged in three rows in *Schizothorax* and in two rows in *Gymnocypris*, *Gymnodiptychus*, *Ptychobarbus*, and *Schizopygopsis*. The dental formula is usually 2.3.5(4)-5(4).3.2 in *Schizothorax* and 3(2).4-4.3(2) in the other genera. All the teeth are Type 2. The central teeth and minor-row teeth are always Type 2-2, and the anterior teeth are usually Type 2-1. The central teeth are never Type 2-3 as in Barbinae.

Schizothorax grahami (Regan, 1904) (Fig. 303-1: IHCAS8260395, BL 117.0 mm)

One specimen from Sichuan, China (IHCAS8260395). Dental formula: 2.3.5-5.3.2; central teeth and minor-row teeth Type 2-2, anterior teeth Type 2-1. Crowns of posterior central and minor-row teeth bent forward with grinding surfaces oriented dorsad. Shallow U-shaped groove on grinding surface of central teeth and minor-row teeth.

Schizothorax lissolabiatus Tsao, 1964 (Fig. 303-2: IHCAS646138, BL 152.0 mm)

One specimen from Yunnan, China (IHCAS646138). Dental formula: 2.3.4-4.3.2; central teeth and minor-row teeth Type 2-2, anterior teeth Type 2-0. Crowns of central teeth and minor-row teeth bent forward with grinding surfaces oriented dorsad. Shallow U-shaped groove on grinding surface of central teeth. Both margins of grinding surface of posterior minor-row teeth jutting out.

Schizothorax meridionalis Tsao, 1964 (Fig. 303-3: IHCAS9040155, BL 176.0 mm)

One specimen from Yunnan, China (IHCAS9040155). Dental formula: 2.3.5-5.3.2; central teeth and minor-row teeth Type 2-2, anterior teeth Type 2-1. Crowns of central teeth and posterior minor-row teeth bent forward with grinding surfaces oriented dorsad. Both margins of grinding surface of posterior minor-row teeth jutting out.

Schizothorax prenanti (Tchang, 1930) (Fig. 303-4: IHCAS877798, BL 148.0 mm)

One specimen from Hubei, China (IHCAS877798). Dental formula: 2.3.5-5.3.2; central teeth and minor-row teeth Type 2-2, anterior teeth Type 2-1 or 2-0. Crowns of posterior central and minor-row teeth bent forward with grinding surfaces oriented dorsad. Fuzzy U-shaped groove on grinding surface of central teeth and minor-row teeth. Crowns of some teeth worn.

Schizothorax wangchiachii (Fang, 1936) (Fig. 303-5: IHCAS8172375, BL 149.0 mm)

One specimen from Yunnan, China (IHCAS8172375). Dental formula: 2.3.5-5.3.2; central teeth and minor-row teeth Type 2-2, anterior teeth Type 2-2 or 2-1. Crowns of posterior central and minor-row teeth bent forward with grinding surfaces oriented dorsad. U-shaped groove on grinding surface of central teeth and minor-row teeth.

Schizothorax labiatus (McClelland, 1842) (Fig. 303-6: IHCAS7681231, BL 148.0 mm)

One specimen from Xizang, China (IHCAS7681231). Dental formula: 2.3.4-5.3.2; central teeth and minor-row teeth Type 2-2, anterior teeth Type 2-2 or 2-0. Crowns of posterior central and minor-row teeth bent forward with grinding surfaces oriented dorsad. Broad and shallow U-shaped groove on grinding surface of central teeth and minor-row teeth. Crowns of some teeth worn.

Schizothorax plagiostomus Heckel, 1838 (Fig. 303-7: IHCAS14054, BL 167.0 mm)

One specimen from Xizang, China (IHCAS no number). Dental formula: -4.3.2; all teeth Type 2-2. Crowns of central teeth and minor-row teeth bent forward with grinding surfaces oriented dorsad. Longitudinal ridge present on grinding surface of posterior central teeth and minor-row teeth.

Schizothorax dolichonema Herzenstein, 1889 (Fig. 303-8: IHCAS0302, BL 112.0 mm)

One specimen from Sichuan, China (IHCAS0302). Dental formula: 2.3.5-4.3.2; all teeth Type 2-2. Crowns of posterior central and minor-row teeth bent forward with grinding surfaces oriented dorsad. Shallow U-shaped groove on grinding surface of central teeth and minor-row teeth. Grinding surface of posterior minor-row teeth wrinkled, with both margins jutting out.

Gymnocypris dobula Günther, 1868 (Fig. 303-9: IHCAS7580421, BL 142.0 mm)

One specimen from Xizang, Chine (IHCAS7580421). Dental formula: 3.4-4.3; central teeth and minor-row teeth Type 2-2, anterior teeth Type 2-1. Crowns of posterior central and minor-row teeth bent forward with grinding surfaces oriented dorsad. Grinding surfaces of posterior central teeth wrinkled. Shallow U-shaped groove on grinding surface of central teeth and minor-row teeth. Some teeth worn.

Gymnocypris eckloni **Herzenstein, 1891** (Fig. 303-10: IHCAS9090154, BL 191.0 mm)

One specimen from Gansu, China (IHCAS9090154). Dental formula: 3.4-4.3; central teeth and minor-row teeth Type 2-2, anterior teeth Type 2-1. Crowns of posterior central and minor-row teeth bent forward with grinding surfaces oriented dorsad. Shallow U-shaped groove on grinding surface of central teeth and minor-row teeth. Grinding surface of posterior central teeth and minor-row teeth wrinkled.

Gymnocypris przewalskii **(Kessler, 1876)** (Fig. 303-11: IHCAS9090248, BL 138.0 mm)

One specimen from Qinhai, China (IHCAS9090248). Dental formula: 3.4-4.3; central teeth and minor-row teeth Type 2-2, anterior teeth Type 2-1. Crowns of central teeth and minor-row teeth bent forward with grinding surfaces oriented dorsad. Shallow U-shaped groove on grinding surface of central teeth and minor-row teeth.

Gymnocypris waddelli **Regan, 1905** (Fig. 303-12: IHCAS7590967, BL 114.0 mm)

One specimen from Xizang, China (IHCAS74590967). Dental formula: 3.4-4.3; central teeth and minor-row teeth Type 2-2, anterior teeth Type 2-1. Crowns of posterior central and minor-row teeth bent forward with grinding surfaces oriented dorsad. U-shaped groove on grinding surface of central teeth and minor-row teeth. Several parallel ridges present on grinding surface of posterior central teeth and minor-row teeth. Some teeth worn.

Gymonocypris chilianensis **Li & Chang, 1974** (Fig. 303-13: IHCAS7772869, BL 104.0 mm)

One specimen from Gansu, China (IHCAS7772869). Dental formula: 3.4-4.3; central teeth and minor-row teeth Type 2-2, anterior teeth Type 2-1. Crowns of posterior central and minor-row teeth bent forward with grinding surfaces oriented dorsad. U-shaped groove on grinding surface of central teeth and minor-row teeth.

Gymnodiptychus pachycheilus **Herzenstein, 1892** (Fig. 303-14: IHCAS8272074, BL 163.0 mm)

Two specimens from Sichuan, China (IHCAS8272074 and 615028). Dental formula: 3.4-4.3; central teeth and minor-row teeth Type 2-2, anterior teeth Type 2-1. Crowns of central teeth and minor-row teeth bent forward with grinding surfaces oriented dorsad.

Ptychobarbus chungtienensis **(Tsao, 1964)** (Fig. 303-15: IHCAS8193331, BL 132.0 mm)

One specimen from Yunnan, China (IHCAS8193331). Dental formula: 3.4-4.3; central teeth and minor-row teeth Type 2-2, anterior teeth Type 2-1. Crowns of central teeth and minor-row teeth bent forward with grinding surfaces oriented dorsad. Crowns of some teeth worn.

Ptychobarbus kaznakovi **Nikolskii, 1903** (Fig. 303-16: IHCAS0780, BL 159.0 mm)

One specimen from Sichuan, China (IHCAS0780). Dental formula: 3.4-4.3; all teeth Type 2-2. Crowns of posterior central and minor-row teeth bent forward with grinding surfaces oriented dorsad. Crowns well worn with secondary grinding surfaces, and compressed antero-posteriorly.

Schizopygopsis malacanthus **Herzenstein, 1891** (Fig. 303-17: IHCAS no number, BL 96.0 mm)

Three specimen from Xizang, China (IHCAS7671785, no numbers). Dental formula: 3.4-4.3; central teeth and minor-row teeth Type 2-2, anterior teeth Type 2-1. Crowns of posterior central and minor-row teeth bent forward with grinding surfaces oriented dorsad. Shallow U-shaped groove on grinding surface of central teeth and minor-row teeth. Several parallel fine ridges present on grinding surface of posterior central teeth and minor-row teeth.

Schizopygopsis pylzovi **Kessler, 1876** (Fig. 303-18: IHCAS no number, BL 136.0 mm)

One specimen from Sichuan, China (IHCAS no number). Dental formula: 3.4-4.3; all teeth Type 2-2. Crowns of central teeth and minor-row teeth bent forward with grinding surfaces oriented dorsad. U-shaped groove on grinding surface of central teeth and minor-row teeth. Crowns of some teeth well worn.

Schizopygopsis stoliczkai **Steindachner, 1866** (Fig. 303–19: IHCAS7660558, BL 232.0 mm)

One specimen from Xizang, China (IHCAS7660558). Dental formula: 3.4–4.3; all teeth Type 2–2. Crowns of posterior minor-row teeth bent forward with grinding surfaces oriented dorsad. U-shaped groove on grinding surface of central teeth and minor-row teeth. Crowns of some teeth well worn with secondary grinding surface developed.

Schizopygopsis younghasbandi **Regan, 1905** (Fig. 303–20: IHCAS7390499, BL 138.0 mm)

One specimen from Xizang, China (IHCAS7390499). Dental formula: 2.4–4.2; central teeth and minor-row teeth Type 2–2, anterior teeth Type 2–1. Crowns of central teeth and minor-row teeth bent forward with grinding surfaces oriented dorsad. Transverse groove on grinding surface of major-row teeth and minor-row teeth. Several fine longitudinal parallel ridges present on grinding surface of posterior central teeth and minor-row teeth.

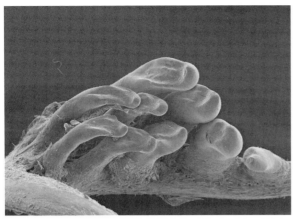

Fig. 303-1. *Schizothorax grahami*: IHCAS8260395.

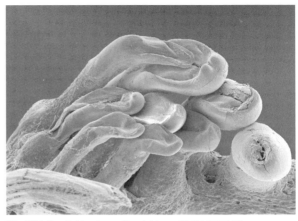

Fig. 303-2. *Schizothorax lissolabiatus*: IHCAS646138.

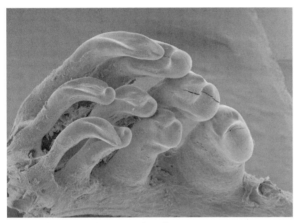

Fig. 303-3. *Schizothorax meridionalis*: IHCAS9040155.

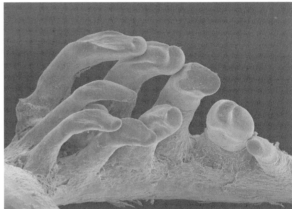

Fig. 303-4. *Schizothorax prenanti*: IHCAS877798.

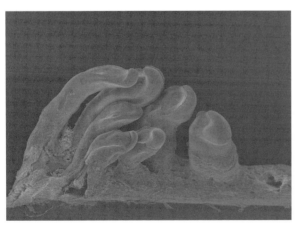

Fig. 303-5. *Schizothorax wangchiachii*: IHCAS8172375.

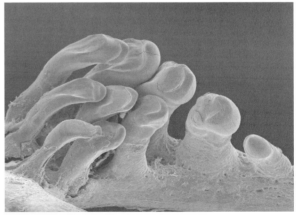

Fig. 303-6. *Schizothorax labiatus*: IHCAS7681231.

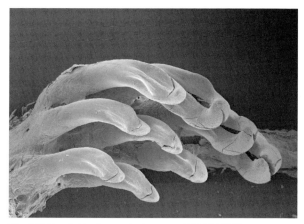

Fig. 303-7. *Schizothorax plagiostomus*: IHCAS14054.

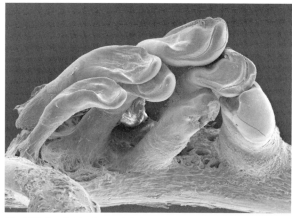

Fig. 303-8. *Schizothorax dolichonema*: IHCAS0302.

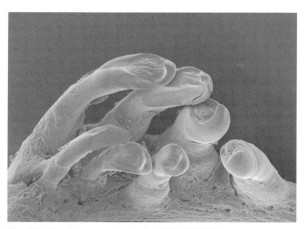

Fig. 303-9. *Gymnocypris dobula*: IHCAS7580421.

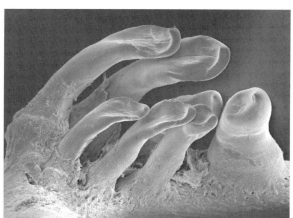

Fig. 303-10. *Gymnocypris eckloni*: IHCAS9090154.

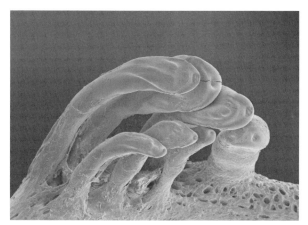

Fig. 303-11. *Gymnocypris przewalskii*: IHCAS9090248.

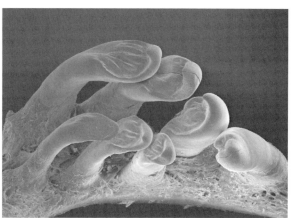

Fig. 303-12. *Gymnocypris waddelli*: IHCAS7590967.

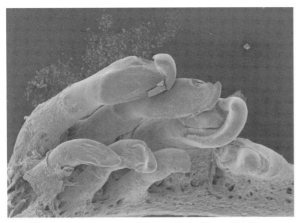

Fig. 303-13. *Gymonocypris chilianensis*: IHCAS7772869.

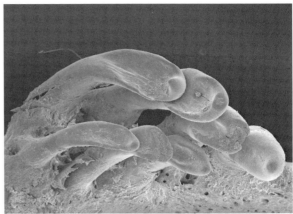

Fig. 303-14. *Gymnodiptychus pachycheilus*: IHCAS8272074.

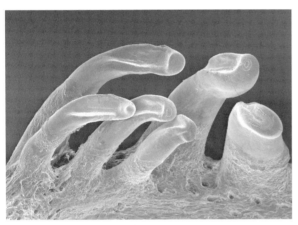

Fig. 303-15. *Ptychobarbus chungtienensis*: IHCAS8193331.

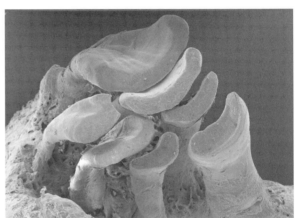

Fig. 303-16. *Ptychobarbus kaznakovi*: IHCAS0780.

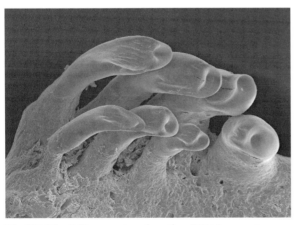

Fig. 303-17. *Schizopygopsis malacanthus*: IHCAS no number.

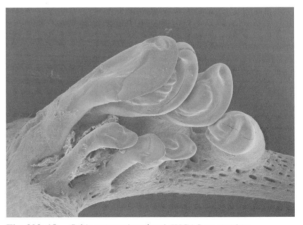

Fig. 303-18. *Schizopygopsis pylzovi*: IHCAS no number.

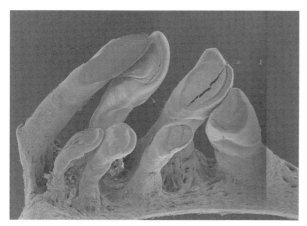 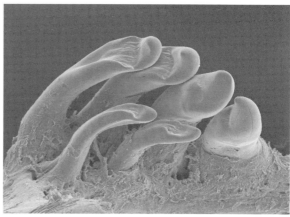

Fig. 303-19. *Schizopygopsis stoliczkai*: IHCAS7660558.   **Fig. 303-20.** *Schizopygopsis younghasbandi*: IHCAS7390499.

3-4. Cultrinae

The pharyngeal dentitions of 25 species in 12 genera of Cultrinae were examined. Their teeth are arranged in three rows. The typical dental formula is 2(1).4(3).5(4)–5(4).4(3).2(1), as in Danioninae. The central teeth are usually Type 3-0 in nearly all species. Hamulus grooves are present on the posterior or medial sides of the crowns of the central teeth and minor-row teeth in almost every species.

Paralaubuca barroni (**Fowler, 1934**) (Fig. 304-1: IHCAS7850302, BL 85.7 mm)

Two specimens from Yunnan, China (IHCAS7850302 and 7850300). Dental formula: 2.4.4–5.4.2; central teeth Type 3-0, anterior teeth Type 2-3 or 2-1, minor-row teeth Type 3-0, 2-3 or 2-1. Crowns of posterior central and minor-row teeth bent forward. Crowns of major-row teeth compressed antero-posteriorly or medio-laterally. Tiny hamulus groove present on posterior or medial side of central teeth and minor-row teeth.

Paralaubuca riveroi (**Fowler, 1935**) (Fig. 304-2: LBM1210047512, BL 106.3 mm)

One specimen from Thailand (LBM1210047512). Dental formula: 2.3.4–4.4.2; central teeth Type 3-0, anterior teeth Type 2-1, minor-row teeth Type 2-3 or 2-1. Crowns of posterior central and minor-row teeth bent forward. Crowns of anterior teeth and anterior minor-row teeth compressed medio-laterally. Tiny hamulus groove present on posterior or medial side of central teeth and minor-row teeth.

Paralaubuca typus **Bleeker, 1864** (Fig. 304-3: LBM1210013846, BL 90.9 mm)

Two specimens from Thailand (LBM1210013845 and 1210013846). Dental formula: 2.4.5–4.4.2; central teeth Type 3-0, anterior teeth Type 2-1, minor-row teeth Type 3-0, 2-3 or 2-1. Crowns of posterior central and minor-row teeth bent forward. Crowns of anterior teeth and anterior minor-row teeth compressed medio-laterally. Tiny hamulus grooves present on posterior or medial side of central teeth and minor-row teeth. Many denticles present on grinding surface of all teeth.

Chanodichthys dabryi (**Bleeker, 1871**) (Fig. 304-4: IHCAS0532, BL 210.5 mm)

Five specimens from Heilongjiang (IHCAS0532

and 0529) and Hubei, China (LBM1210013740, IHCAS79110060, and 79110062). Dental formula: 2.4(3).5(4)–5(4).4(3).2; central teeth Type 3-0 or 2-3, anterior teeth Type 2-1, minor-row teeth Type 3-0, 2-3 or 2-1. Tooth crowns compressed medio-laterally or antero-posteriorly. Tiny hamulus groove present on posterior or medial side of crown of central teeth and minor-row teeth.

Chanodichthys erythropterus (**Basilewsky, 1855**) (Fig. 304-5: IHCAS580906, BL 253.5 mm)

Fourteen specimens from Heilongjiang, (IHCAS580906) and Hubei, China (LBM1210038429–1210038441). Dental formula: 2(1).4(3).5(4)–5(4).4.2(1); central teeth Type 3-0 or 2-3, anterior teeth Type 2-1, minor-row teeth Type 3-0, 2-3 or 2-1. Crowns of posterior central and minor-row teeth bent forward. Tiny hamulus groove present on posterior or medial side of crown of central teeth and minor-row teeth.

Chanodichthys mongolicus (**Basilewsky, 1855**) (Fig. 304-6: LBM1210013828, BL 300.0 mm)

Five specimens from Hubei (LBM1210013827 and 1210013828) and Sichuan, China (IHCAS90953, 586024, and 586463). Dental formula: 2.4.5(4)–5(4).4.2; central teeth Type 3-0 or 2-3, anterior teeth Type 2-1, minor-row teeth Type 2-3 or 2-1. Crowns of posterior central teeth bent forward. Crowns of anterior teeth and anterior minor-row teeth compressed medio-laterally. Tiny hamulus groove present on posterior or medial side of crown of central teeth and minor-row teeth.

Culter alburnus **Basilewsky, 1855** (Fig. 304-7: LBM1210013830, BL 240.0 mm)

Five specimens from Hubei (LBM1210013829, 1210013830, 1210013831), Zhejiang, (LBM1210048360), and Heilongjiang, China (IHCAS8990186). Dental formula: 2(1).4.5(4)–5(4).4.2; central teeth Type 3-0, anterior teeth Type 2-1, minor-row teeth Type 3-0, 2-3 or 2-1. Crowns of posterior central and minor-row teeth bent forward. Crowns of anterior teeth and anterior minor-row teeth compressed medio-laterally. Tiny hamulus groove present on posterior or medial side of crown of central teeth and minor-row teeth.

Culter oxicephaloides **Kreyenberg & Pappenheim, 1908** (Fig. 304-8: IHCAS6491904, BL 148.9 mm)

Three specimens from Hunan, China

(IHCAS6350743, 6491904, and 6491903). Dental formula: 1(2).4.5(4)‒5(4).4.2(1); central teeth Type 3‒0, anterior teeth Type 2‒1, minor-row teeth Type 3‒0, 2‒3 or 2‒1. Crowns of posterior central and minor-row teeth bent forward. Crowns of anterior teeth and anterior minor-row teeth compressed medio-laterally. Tiny hamulus groove present on posterior or medial side of crown of central teeth and minor-row teeth.

Culter recurviceps (Richardson, 1846) (Fig. 304‒9: IHCAS7654978, BL 182.0 mm)

One specimen from Guandong, China (IHCAS7654978). Dental formula: 2.4.5‒4.4.2; central teeth Type 3‒0, anterior teeth and minor-row teeth Type 2‒3 or 2‒1. Crowns of posterior central and minor-row teeth bent forward. Crowns of anterior teeth and anterior minor-row teeth compressed medio-laterally. Tiny hamulus groove present on posterior or medial side of crown of central teeth and minor-row teeth.

Ancherythroculter lini Luo, 1994 (Fig. 304‒10: IHCAS7541126, BL 149.1 mm)

One specimen from Guangxi, China (IHCAS7541126). Dental formula: 2.4.5‒4.4.2; central teeth Type 3‒0, anterior teeth Type 2‒1, minor-row teeth Type 2‒3 or 2‒1. Crowns of posterior central and minor-row teeth bent forward. Crowns of anterior teeth and anterior minor-row teeth compressed medio-laterally. Tiny hamulus groove present on posterior or medial side of crown of central teeth and minor-row teeth.

Hemiculterella sauvagei Warpachowski, 1887 (Fig. 304‒11: IHCAS8840960, BL 105.2 mm)

One specimen from Guizhou, China (IHCAS8840960). Dental formula: 2.4.5‒5.3.2; central teeth Type 3‒0, anterior teeth 2‒3 or 2‒1, minor-row teeth Type 2‒1. Crowns of posterior central and minor-row teeth bent forward. Crowns of anterior teeth and anterior minor-row teeth compressed medio-laterally. Tiny hamulus groove present on posterior or medial side of crown of central teeth and minor-row teeth.

Hemiculterella wui (Wang, 1935) (Fig. 304‒12: IHCAS8130702, BL 125.1 mm)

Four specimens from Guangxi (IHCAS8608262, 8608263, and 86082625) and Anhui, China (IHCAS8130702). Dental formula: 2(1).4.5(4)‒5(4).4.2; central teeth Type 3‒0, anterior teeth Type 2‒1, minor-row

teeth Type 2‒3 or 2‒1. Crowns of posterior central and minor-row teeth bent forward. Crowns of anterior teeth and anterior minor-row teeth compressed medio-laterally. Tiny hamulus groove present on posterior or medial sidesof crows of central teeth and minor-row teeth.

Anabarilius grahami (Regan, 1908) (Fig. 304‒13: IHCAS636553, BL 115.4 mm)

Two specimens from Yunnan, China (IHCAS636553). Dental formula: (1,2).(3,4).(4,5)‒5.4.(1,2); central teeth Type 3‒0, anterior teeth and minor-row teeth Type 2‒3 or 2‒1. Crowns of posterior central and minor-row teeth bent forward. Crowns of anterior teeth and anterior minor-row teeth compressed medio-laterally. Small hamulus groove present on posterior or medial side of crown of central teeth and minor-row teeth.

Anabarilius macrolepis Yih & Wu, 1964 (Fig. 304‒14: IHCAS578084, BL 70.5 mm)

Two specimens from Yunnan, China (IHCAS578033 and 578084). Dental forumula: 2(1).4(3).5‒5.4(3).2; central teeth Type 3‒0, anterior teeth and minor-row teeth Type 2‒3 or 2‒1. Crowns of posterior central and minor-row teeth bent forward. Crowns of anterior teeth and anterior minor-row teeth compressed medio-laterally. Small hamulus groove present on posterior or medial side of crown of central teeth and minor-row teeth.

Anabarilius polylepis (Regan, 1904) (Fig. 304‒15: IHCAS634228, BL 166.1 mm)

Three specimens from Yunnan, China (IHCAS634228, 634235, and 634340). Dental formula: 2(1).4.5(4)‒5(4).4.2(1); central teeth Type 3‒0, anterior teeth Type 3‒0 or 2‒3, minor-row teeth Type 2‒3 or 2‒1. Crowns of posterior central and minor-row teeth bent forward. Crowns of anterior teeth and anterior minor-row teeth compressed medio-laterally. Small hamulus groove present on posterior or medial side of crown of central teeth and minor-row teeth.

Anabarilius transmontanus (Nichols, 1925) (Fig. 304‒16: IHCAS858474, BL 140.7 mm)

Two specimens from Yunnan, China (IHCAS858474 and 858479). Dental formula: 2.4.(4,5)‒(4,5).(3,4).2; central teeth Type 3‒0, anterior teeth and minor-row teeth Type 2‒3 or 2‒1. Crowns of posterior central and minor-row teeth bent forward. Crowns of anterior teeth and anterior minor-row teeth compressed medio-laterally.

3. Descriptions of the pharyngeal dentition of cyprinid subfamilies **81**

Small hamulus groove present on posterior or medial side of crown of central teeth and minor-row teeth.

Hemiculter leucisculus (Basilewsky, 1855) (Fig. 304‒17: LBM1210015901, BL 230.0 mm)

Fourteen specimens from Hubei, China (LBM1210014580, 1210014965, 1210015901‒1210015909, 1210016131, 1210016132, and 1210036534). Dental formula: 2(1).4(3).5(4)‒5(4).4(3).2(1); central teeth Type 3‒0, anterior teeth Type 2‒1, minor-row teeth Type 2‒3 or 2‒1. Crowns of posterior central and minor-row teeth bent forward. Crowns of anterior teeth compressed medio-laterally. Small hamulus groove present on posterior or medial side of crown of central teeth and minor-row teeth.

Hemiculter lucidus (Dybowski, 1872) (Fig. 304‒18: LBM1210047437, BL 143.5 mm)

Two specimens from Zhejiang, China (LBM1210047437 and 1210047439). Dental formula: 2.4.5‒4.4.2; central teeth Type 3‒0, anterior teeth Type 2‒1, minor-row teeth Type 2‒3 or 2‒1. Crowns of posterior central and minor-row teeth bent forward. Crowns of anterior teeth compressed medio-laterally. Small hamulus groove present on posterior or medial side of crown of central teeth and minor-row teeth.

Ischikauia steenackeri (Sauvage, 1883) (Fig. 304‒19: LBM1210003655, BL 86.1 mm)

Twelve specimens from Shiga, Japan (LBM1210003654‒1210003666, 1210014591, 1210014592, 1210014600, and 1210014601). Dental formula: 2.4.5(4)‒5(4).4.2; central teeth Type 3‒0, anterior teeth and minor-row teeth Type 2‒3 or 2‒1. Crowns of posterior central and minor-row teeth bent forward. Crowns of major-row teeth and anterior minor-row teeth compressed antero-posteriorly or medio-laterally. Hamulus groove present on posterior or medial side of crown of central teeth, minor-row teeth, and tooth A2 (An2).

Megalobrama amblycephala Yih, 1955 (Fig. 304‒20: LBM1210014587, BL 78.3 mm)

Seven specimens from Hubei, China (IHCAS570447 and 550013, and LBM1210013824, 1210014585‒1210014587, and 1210017720). Dental formula: 2.4.5(4)‒5(4).4(3).2(1); central teeth Type 3‒0, anterior teeth Type 2‒3 or 2‒1, minor-row teeth Type 3‒0, 2‒3 or 2‒1. Crowns of posterior central and minor-row teeth bent forward. Crowns of major-row teeth and anterior minor-row teeth

compressed antero-posteriorly or medio-laterally. Large hamulus groove present on posterior or medial side of crown of central teeth, minor-row teeth, and tooth A2 (An2).

Megalobrama terminalis (Richardson, 1846) (Fig. 304‒21: IHCAS5469, BL 147.2 mm)

Three specimens from Hunan, China (IHCAS5469 and 5431, and LBM1210015022). Dental formula: 2.4(3).5‒5.4.2; central teeth Type 3‒0, anterior teeth and minor-row teeth Type 2‒3 or 2‒1. Crowns of posterior central and minor-row teeth bent forward. Crowns of major-row teeth and anterior minor-row teeth compressed antero-posteriorly or medio-laterally. Large hamulus groove present on posterior or medial side of crown of central teeth, minor-row teeth, and tooth A2 (An2).

Metzia lineata (Pellegrin, 1907) (Fig. 304‒22: IHCAS83100981, BL 75.7 mm)

One specimen from Guangxi, China (IHCAS83100981). Dental formula: 2.3.4‒4.4.2; central teeth Type 3‒0, anterior teeth Type 2‒3 or 2‒1, minor-row teeth Type 3‒0, 2‒3 or 2‒1. Crowns of posterior central and minor-row teeth bent forward. Crowns of anterior teeth compressed medio-laterally. Large hamulus groove present on posterior or medial side of crown of central teeth, minor-row teeth, and tooth A2 (An2).

Parabramis pekinensis (Basilewsky, 1855) (Fig. 304‒23: LBM1210013825, BL 270.0 mm)

Two specimens from Hubei, China (LBM1210013825 and 1210013826). Dental formula: 2.4.5‒(4,5).4.2; central teeth Type 3‒0, anterior teeth Type 2‒3 or 2‒1, minor-row teeth Type 3‒0, 2‒3 or 2‒1. Crowns of posterior central and minor-row teeth bent forward. Crowns of major-row teeth and anterior minor-row teeth compressed medio-laterally or antero-posteriorly. Large hamulus groove is present on posterior side of all tooth crown in all teeth.

Sinibrama macrops (Günther, 1868) (Fig. 304‒24: IHCAS0073, BL 132.9 mm)

Three specimens from Zhejinag, China (IHCAS0073, 0075, and 0077). Dental formula: 2.4.4(5)‒4.4.2; central teeth Type 3‒0, anterior teeth Type 2‒1, minor-row teeth Type 3‒0, 2‒3 or 2‒1. Crowns of posterior central and minor-row teeth bent forward. Crowns of major-row teeth and anterior minor-row teeth compressed medio-laterally or antero-posteriorly. Large hamulus groove present on posterior or medial side of crown of central teeth and minor-row teeth.

Sinibrama melrosei **(Nichols & Pope, 1927)** (Fig. 304–25: IHCAS7553158, BL 111.6 mm)

Three specimens from Guangxi, China (IHCAS7553158, 7541140, and 7553258). Dental formula: 2.4.4–4(5).4.2; central teeth Type 3–0, anterior teeth and minor-row teeth Type 2–3 or 2–1. Crowns of posterior central and minor-row teeth bent forward. Crowns of major-row teeth and anterior minor-row teeth compressed medio-laterally or antero-posteriorly. Large hamulus groove present on posterior or medial side of crown of central teeth and minor-row teeth.

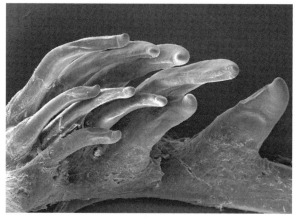

Fig. 304-1. *Paralaubuca barroni*: IHCAS7850302.

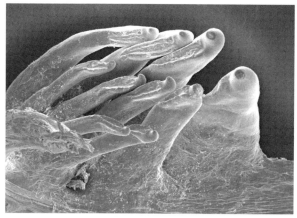

Fig. 304-2. *Paralaubuca riveroi*: LBM1210047512.

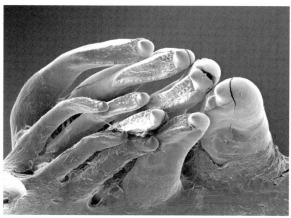

Fig. 304-3. *Paralaubuca typus*: LBM1210013846.

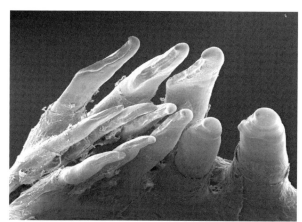

Fig. 304-4. *Chanodichthys dabryi*: IHCAS0532.

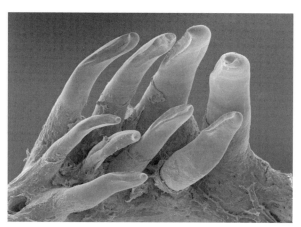

Fig. 304-5. *Chanodichthys erythropterus*: IHCAS580906.

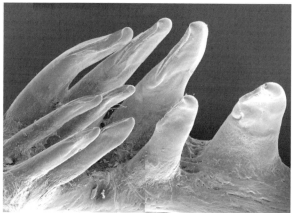

Fig. 304-6. *Chanodichthys mongolicus*: LBM1210013828.

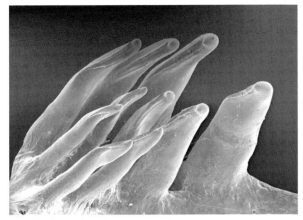

Fig. 304-7. *Culter alburnus*: LBM1210013830.

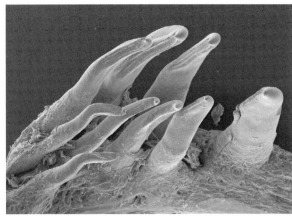

Fig. 304-8. *Culter oxicephaloides*: IHCAS6491904.

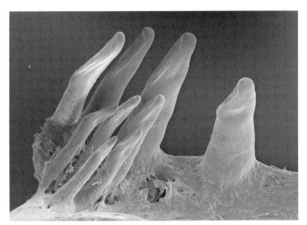

Fig. 304-9. *Culter recurviceps*: IHCAS7654978.

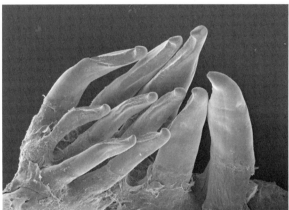

Fig. 304-10. *Ancherythroculter lini*: IHCAS7541126.

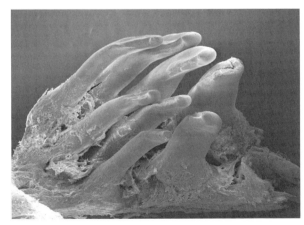

Fig. 304-11. *Hemiculterella sauvagei*: IHCAS8840960.

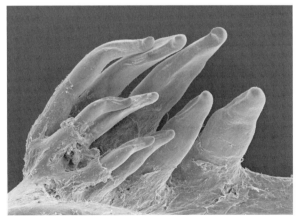

Fig. 304-12. *Hemiculterella wui*: IHCAS8130702.

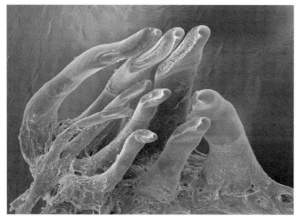

Fig. 304−13. *Anabarilius grahami*: IHCAS636553.

Fig. 304−14. *Anabarilius macrolepis*: IHCAS578084.

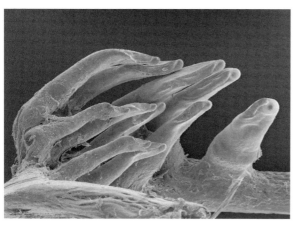

Fig. 304−15. *Anabarilius polylepis*: IHCAS634228.

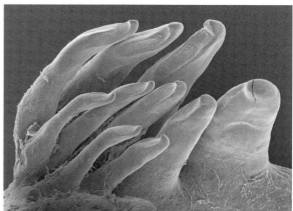

Fig. 304−16. *Anabarilius transmontanus*: IHCAS858474.

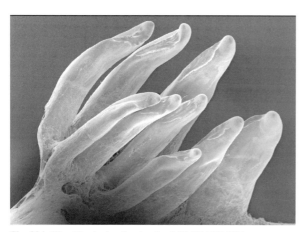

Fig. 304−17. *Hemiculter leucisculus*: LBM1210015901.

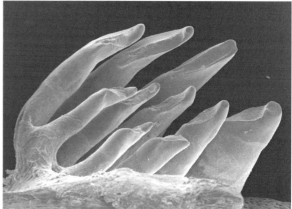

Fig. 304−18. *Hemiculter lucidus*: LBM1210047437.

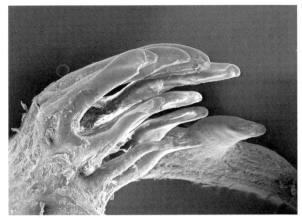

Fig. 304-19. *Ischikauia steenackeri*: LBM1210003655.

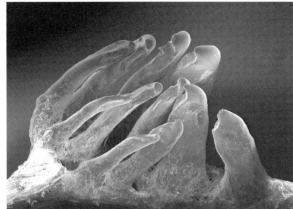

Fig. 304-20. *Megalobrama amblycephala*: LBM1210014587.

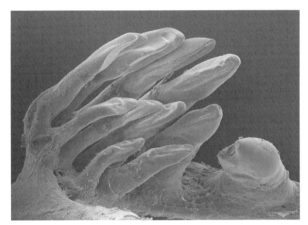

Fig. 304-21. *Megalobrama terminalis*: IHCAS5469.

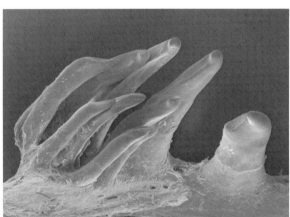

Fig. 304-22. *Metzia lineata*: IHCAS83100981.

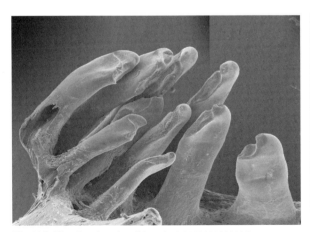

Fig. 304-23. *Parabramis pekinensis*: LBM1210013825.

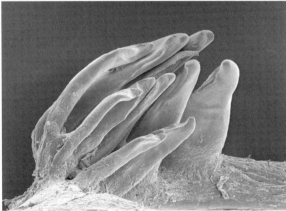

Fig. 304-24. *Sinibrama macrops*: IHCAS0073.

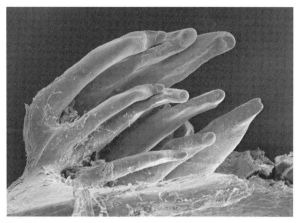

Fig. 304-25. *Sinibrama melrosei*: IHCAS7553158.

3-5. Xenocypridinae

The pharyngeal dentitions of five species in four genera of Xenocypridinae were examined. Their teeth are arranged in one to three rows. There are three rows in the genera *Xenocypris* and *Plagiognathops*, two rows in *Distoechodon*, and one row in *Pseudobrama*. The six to seven major-row teeth of xenocypridines exceed those of general cyprinids in number. The general dental formula is 2.4.6(7)‒6(7).4.2 in *Xenocypris* and *Plagiognathops*, 3(4).6‒6.3(4) in *Distoechodon*, and 6‒6 in *Pseudobrama*. All the teeth are Type 3‒0. Their tooth crowns are intensely worn, and secondary grinding surfaces are formed. The major-row teeth are compressed antero-posteriorly, and the minor-row teeth are slender and rod-like.

***Xenocypris davidi* Bleeker, 1871** (Fig. 305‒1: IHCAS no number, BL 298.6 mm)

One specimen from Hubei, China (IHCAS no number). Dental formula: 2.4.6‒6.4.2; all teeth Type 3‒0. Crowns of major-row teeth compressed antero-posterorly. Tooth crowns well worn with secondary grinding surfaces. Minor-row teeth slender and rod-like.

***Xenocypris macrolepis* Bleeker, 1871** (Fig. 305‒2: IHCAS no number, BL 104.0 mm)

One specimen from Hubei, China (IHCAS no number). Dental formula: 2.4.6‒7.4.2; all teeth Type 3‒0. Crowns of major-row teeth compressed antero-posterorly. Tooth crowns well worn with secondary grinding surfaces. Minor-row teeth slender and rod-like.

***Plagiognathops microlepis* (Bleeker, 1871)** (Fig. 305‒3: LBM1210013957, BL 300.0 mm)

Three specimens from Hubei, China (LBM1210013957, 1210013958, and 1210023596). Dental formula: 2.4(2).6(7)‒6.4.2; all teeth Type 3‒0. Crowns of major-row teeth compressed antero-posterorly. Tooth crowns well worn with secondary grinding surfaces. Minor-row teeth slender and rod-like.

***Distoechodon hupeinensis* (Yih, 1964)** (Fig. 305‒4: LBM1210047422, BL 94.0 mm)

Three specimens from Hubei, China (LBM1210047420‒1210047422). Dental formula: 3(4).6‒6.3(4); all teeth Type 3‒0. Crowns of major-row teeth compressed antero-posteriorly. Tooth crowns well worn with secondary grinding surfaces. Minor-row teeth slender and rod-like.

***Pseudobrama simoni* (Bleeker, 1864)** (Fig. 305‒5: LBM1210026913, BL 67.0 mm)

One specimen from Hubei, China (LBM1210026913). Dental formula: 6‒6; all teeth Type 3‒0. Crowns of major-row teeth compressed antero-posterorly. Tooth crowns well worn with secondary grinding surfaces.

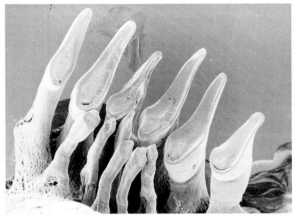

Fig. 305-1. *Xenocypris davidi*: IHCAS no number.

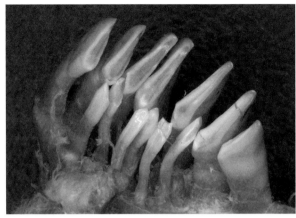

Fig. 305-2. *Xenocypris macrolepis*: IHCAS no number.

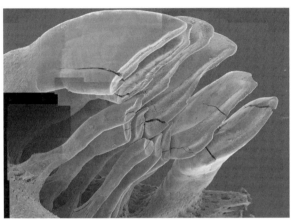

Fig. 305-3. *Plagiognathops microlepis*: LBM1210013957.

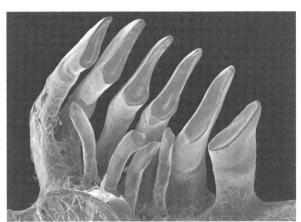

Fig. 305-4. *Distoechodon hupeinensis*: LBM1210047422.

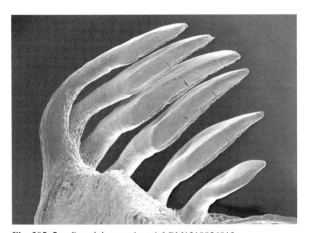

Fig. 305-5. *Pseudobrama simoni*: LBM1210026913.

3-6. Hypophthalmichthyinae

The dentitions of two species in one genus of Hypophthalmichthyinae were examined. The teeth are arranged in one row. The dental formula is 4-4. All the teeth are 3-1, depressed and somewhat slipper-shaped with a finely roughened, tread-like grinding surface. No variation in tooth number was appearent in examined specimens.

Hypophthalmichthys molitrix (Valenciennes, 1844)
(Fig. 306-1: LBM1210013805, BL 238.0 mm)

Eleven specimens from Saitama, Japan (LBM1210013757, 1210013804-1210013808, 1210013822, 1210013898, and 1210013899). Dental formula: 4-4; all teeth Type 3-1. Teeth depressed and somewhat slipper-shaped with a finely roughened, tread-like grinding surface.

Hypophthalmichthys nobilis (Richardson, 1845) (Fig. 306-2: LBM1210013900, BL 132.8 mm)

Nine specimens from Saitama (LBM1210013647, 1210013759, and 1210013900) and Shiga, Japan (LBM1210013619, 1210013646, 1210013659, 1210013672, 1210040519, and 1210040518). Dental formula: 4-4; all teeth Type 3-1. Teeth depressed and somewhat slipper-shaped with grinding surface that appear somewhat fuzzier than that of *H. molitrix*.

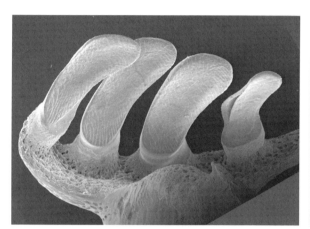

Fig. 306-1. *Hypophthalmichthys molitrix*: LBM1210013805.

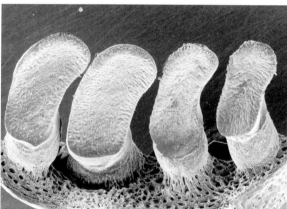

Fig. 306-2. *Hypophthalmichthys nobilis*: LBM1210013900.

3-7. Achielognathinae

The pharyngeal dentitions of 25 species in three genera of Acheilognathinae were examined. Their teeth are arranged in one row, and the dental formula is 5-5 without exception. All the central teeth are Type 3-0 and compressed antero-posteriorly. Several denticles are present on each tooth's anterior margin, which is serrate in most species. Hamulus grooves are present on the posterior sides of the crowns of teeth A2 (An2) to A5 (Po1) in all species. The tooth crowns are worn and secondary grinding surfaces are frequently formed.

***Tanakia tanago* (Tanaka, 1909)** (Fig. 307-1: LBM1210015039, BL 55.8 mm)

Four specimens from Saitama (LBM1210015039 and 1210015040) and Kanagawa, Japan (LBM1210039882 and 1210039884). Dental formula: 5-5; central teeth Type 3-0, anterior teeth Type 3-0 or 2-3. Posterior central teeth bent forward. Tooth crowns compressed antero-posteriorly. Small hamulus groove present on posterior side of crown of teeth A2 (An2) to A5 (Po1). Few denticles present on anterior margin of grinding surface of teeth A2 (An2) to A5 (Po1). Tooth crowns worn.

***Tanakia laceolata* (Temminck and Schlegel, 1846)** (Fig. 307-2: LBM1210014340, BL 68.3 mm)

Eight specimens from Shiga, Japan (LBM121003709, 121003710, 1210014340-1210014343, 1210039876, 1210048590). Dental formula: 5-5; central teeth Type 3-0, anterior teeth Type 3-0 or 2-3. Posterior central teeth bent forward. Tooth crowns compressed antero-posteriorly. Hamulus grooves present on posterior side of crown of teeth A2 (An2) to A5 (Po1). Tooth crowns well worn with secondary grinding surface.

***Tanakia limbata* (Temminck and Schlegel, 1846)** (Fig. 307-3: LBM1210015007, BL 63.1 mm)

Eight specimens from Shiga, Japan (LBM1210003714, 1210015007, 1210015008, 1210043148, 1210043150, 1210043152, 1210043154, and 1210043156). Dental formula: 5-5; central teeth Type 3-0, anterior teeth Type 3-0 or 2-3. Tooth crowns compressed antero-posteriorly. Small hamulus groove present on posterior side of crown of teeth teeth A2 (An2) to A5 (Po1). One or two of large denticles present on anterior margins of grinding surfaces of teeth A2 (An2) to A5 (Po1). Tooth crowns worn.

***Tanakia himantegus* (Günther, 1868)** (Fig. 307-4: LBM1210015036, BL 35.6 mm)

Three specimens from Taiwan, China (LBM1210015036, 1210026890, and 1210026930). Dental formula: 5-5; all teeth Type 3-0. Tooth crowns compressed antero-posteriorly. Small hamulus groove present on posterior side of each tooth crown. Several large denticles separated by diagonal grooves present on anterior margin. Tooth crowns worn.

***Rhodeus atremius* (Jordan and Thompson, 1914)** (Fig. 307-5: LBM1210040405, BL 45.1 mm)

Sixteen specimens from Japan (LBM1210040375-1210040389 and 1210040405). Dental formula: 5-5; all teeth Type 3-0. Tooth crowns compressed antero-posteriorly. Tiny hamulus groove present on posterior side of crown of teeth A2 (An2) to A5 (Po1). Some denticles present on anterior margin of grinding surface of teeth A2 (An2) to A5 (Po1). Tooth crowns well worn.

***Rhodeus suigensis* (Mori, 1935)** (Fig. 307-6: LBM1210040416, BL 42.1 mm)

Seven specimens from Okayama, Japan (LBM1210015173, 1210015174, 1210039833, 1210039835, 1210039837, 1210039839, and 1210039841). Dental formula: 5-5; all teeth Type 3-0. Tooth crowns compressed antero-posteriorly. Tiny hamulus groove present on the posterior side of crown of teeth A2 (An2) to A5 (Po1). Few denticles present on anterior margin of grinding surface of teeth A2 (An2) to A5 (Po1). Tooth crowns well worn with secondary grinding surface.

***Rhodeus fangi* (Miao, 1934)** (Fig. 307-7: LBM1210047475, BL 52.1 mm)

One specimen from Hubei, China (LBM1210047475). Dental formula: 5-5; all teeth Type 3-0. Tooth crowns compressed antero-posteriorly. Small hamulus groove present on posterior side of crown of teeth A2 (An2) to A5 (Po1). Some large denticles separated by diagonal grooves present on anterior margin of grinding surface of teeth A2 (An2) to A5 (Po1). Tooth crowns worn.

***Rhodeus ocellatus* (Kner, 1866)** (Fig. 307-8: LBM1210014636, BL 51.6 mm)

Three specimens from Gifu (LBM1210013761) and Shiga, Japan (LBM1210014636 and 1210014638). Dental formula: 5-5; all teeth Type 3-0. Tooth crowns compressed antero-posteriorly. Small hamulus groove present on posterior side of all tooth crowns. A few denticles present on anterior margin of grinding surface of all tooth crowns. Tooth crowns well worn, with secondary grinding surface.

Rhodeus amarus (**Bloch, 1782**) (Fig. 307-9: LBM1210024448, BL 56.0 mm)

Three specimens from Europe (LBM1210024444, 1210024446, and 1210024448). Dental formula: 5-5; all teeth Type 3-0. Tooth crowns compressed antero-posteriorly. Hamulus groove present on posterior side of all tooth crowns. Denticles present on anterior margin of grinding surface of all teeth. Tooth crowns worn.

Rhodeus sinensis **Günther, 1868** (Fig. 307-10: LBM1210014963, BL 52.2 mm)

One specimen from Korea (LBM1210014963). Dental formula: 5-5; all teeth Type 3-0. Tooth crowns compressed antero-posteriorly. Hamulus groove present on the posterior side of all tooth crowns, large denticles or serrations present on anterior margins, separated by diagonal grooves. Tooth crowns worn.

Acanthorhodeus chankaensis (**Dybowski, 1872**) (Fig. 307-11: LBM1210014960, BL 70.0 mm)

One specimen from Hubei, China (LBM1210014960). Dental formula: 5-5; all teeth Type 3-0. Tooth crowns compressed antero-posteriorly. Hamulus groove present on posterior side of all tooth crowns. Large denticles separated by diagonal grooves present on anterior margin of each tooth's grinding surface. Tooth crowns worn.

Acheilognathus intermedia (**Temminck & Schlegel, 1846**) (Fig. 307-12: LBM1210014993, BL 51.6 mm)

Four specimens from Korea (LBM1210014993-1210014996). Dental formula: 5-5; all teeth Type 3-0. Tooth crowns compressed antero-posteriorly. Small hamulus groove present on posterior side of tooth crown of teeth A2 (An2) to A5 (Po1). Denticles present on anterior margin of grinding surface of teeth A2 to A5. Tooth crowns worn.

Acheilognathus cyanostigma **Jordan & Fowler, 1903** (Fig. 307-13: LBM1210014969, BL 67.5 mm)

Five specimens from Kyoto (LBM1210014968 and 1210014969) and from Shiga, Japan (LBM1210003708, 1210039880, and 1210040511). Dental formula: 5-5; all teeth Type 3-0. Tooth crowns compressed antero-

posteriorly. Hamulus groove present on posterior side of all teeth. Large denticles separated by diagonal grooves present on anterior margin of each tooth's grinding surface. Tooth crowns worn.

Acheilognathus longipinnis **Regan, 1905** (Fig. 307-14: LBM1210015015, BL 64.7 mm)

Four specimens from Osaka, Japan (LBM1210015013-1210015016). Dental formula: 5-5; all teeth Type 3-0. Tooth crowns compressed antero-posteriorly. Hamulus groove present on posterior side of all tooth crowns. Large denticles separated by diagonal grooves present on anterior margin of each tooth's grinding surface. Tooth crowns worn.

Acheilognathus macropterus (**Bleeker, 1871**) (Fig. 307-15: LBM1210014663, BL 83.3 mm)

Two specimens from Jiangsu, China (LBM1210014663 and 1210014664). Dental formula: 5-5; all teeth Type 3-0. Tooth crowns compressed antero-posteriorly. Hamulus groove present on posterior side of all tooth crowns. Large denticles separated by diagonal grooves present on anterior margin of each tooth's grinding surface. Tooth crowns worn.

Acheilognathus melanogaster **Bleeker, 1860** (Fig. 307-16: LBM1210015169, BL 75.8 mm)

Seven specimens from Ibaragi (LBM1210015169, 1210015170, 1210039864, and 1210039866) and from Aomori, Japan (LBM1210040307, 1210040309, and 1210040311). Dental formula: 5-5; all teeth Type 3-0. Tooth crowns compressed antero-posteriorly. Hamulus groove present on posterior side of all tooth crowns. Large denticles separated by diagonal grooves present on anterior margin of each tooth's grinding surface. Tooth crowns worn.

Acheilognathus rhombeus (**Temminck & Schlegel, 1846**) (Fig. 307-17: LBM1210031352, BL 73.1 mm)

Fourteen specimens from Shiga, Japan (LBM1210031328, 1210031330, 1210031332, 1210031334, 1210031336, 1210031338, 1210031340, 1210031342, 1210031344, 1210031346, 1210031348, 1210031350, 1210031352, and 1210031354). Dental formula: 5-5; all teeth Type 3-0. Tooth crowns compressed antero-posteriorly. Hamulus groove present on posterior side of all tooth crowns. Large denticles separated by diagonal grooves present on anterior margin of each tooth's grinding surface. Tooth crowns worn.

Acheilognathus tabira erythropterus **Arai, Fujikawa and Nagata, 2007** (Fig. 307-18: LBM1210015166, BL 70.3 mm)

Five specimens from Ibaragi, Japan (LBM1210015165, 1210015166, 1210040417, 1210041415, and 1210041416). Dental formula: 5-5; all teeth Type 3-0. Tooth crowns compressed antero-posteriorly. Hamulus groove present on posterior side of all tooth crowns. Large denticles separated by diagonal grooves present on anterior margin of each tooth's grinding surface. Tooth crowns worn.

Acheilognathus tabira jordani **Arai, Fujikawa and Nagata, 2007** (Fig. 307-19: LBM1210047474, BL 35.3 mm)

Four specimens from Toyama, Japan (LBM1210047441, 1210047470, 1210047472, and 1210047474). Dental formula: 5-5; all teeth Type 3-0. Tooth crowns compressed antero-posteriorly. Hamulus groove present on posterior side of all tooth crowns. Large denticles separated by diagonal grooves present on anterior margin of each tooth's grinding surface. Tooth crowns worn.

Acheilognathus tabira nakamurae **Arai, Fujikawa and Nagata, 2007** (Fig. 307-20: LBM1210015164, BL 49.2 mm)

Three specimens from Fukuoka, Japan (LBM1210015163, 1210015164, and 1210041487). Dental formula: 5-5; all teeth Type 3-0. Tooth crowns compressed antero-posteriorly. Hamulus groove present on posterior side of all tooth crowns. Large denticles separated by diagonal grooves present on anterior margin of each tooth's grinding surface. Tooth crowns worn.

Acheilognathus tabira tabira **Jordan & Thompson, 1914** (Fig. 307-21: LBM1210024465, BL 67.0 mm)

Eight specimens from Shiga, Japan (LBM1210003712, 1210003713, 1210013773, 1210013815, 1210013948, 1210024279, 1210024465, and 1210039851). Dental formula: 5-5; all teeth Type 3-0. Tooth crowns compressed antero-posteriorly. Hamulus groove present on posterior side of all tooth crowns. Large denticles separated by diagonal grooves present on anterior margin of each tooth's grinding surface. Tooth crowns well worn.

Acheilognathus tabira tohokuensis **Arai, Fujikawa & Nagata, 2007** (Fig. 307-22: LBM1210040437, BL 49.5 mm)

Ten specimens from Akita, Japan (LBM1210040419, 1210040421, 1210040423, 1210040425, 1210040427, 1210040429, 1210040431, 1210040433, 1210040435, and 1210040437). Dental formula: 5-5; all teeth Type 3-0. Tooth crowns compressed antero-posteriorly. Hamulus groove present on posterior side of all tooth crowns. Large denticles present on anterior margins of grinding surfaces. Tooth crowns well worn.

Acheilognathus tonkinensis **(Vaillant, 1892)** (Fig. 307-23: LBM1210014660, BL 94.6 mm)

Four specimens from Jiangsu, China (LBM1210014658-1210014661). Dental formula: 5-5; all teeth Type 3-0. Tooth crowns compressed antero-posteriorly. Hamulus groove present on posterior sides of all tooth crowns. Large denticles separated by transverse grooves present on anterior margin of each tooth's grinding surface. Tooth crowns worn.

Acheilognathus typus **(Bleeker, 1863)** (Fig. 307-24: LBM1210041993, BL mm)

Five specimens from Ibaragi, Japan (LBM1210041989, 1210041991, 1210041993, 1210041995, and 1210041997). Dental formula: 5-5; all teeth Type 3-0. Tooth crowns compressed antero-posteriorly. Hamulus groove present on posterior side of all tooth crowns. Large denticles separated by diagonal grooves present on anterior margin of each tooth's grinding surface. Tooth crowns worn.

Acheilognathus yamatsutae **Mori, 1928** (Fig. 307-25: LBM1210014962, BL 77.0 mm)

Three specimens from Korea (LBM1210014962, 1210048588, and 1210048589). Dental formula: 5-5; all teeth Type 3-0. Tooth crowns compressed antero-posteriorly. Hamulus groove present on posterior side of all tooth crowns. Large denticles present on anterior margin of grinding surface. Tooth crowns worn.

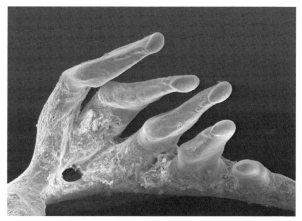

Fig. 307⁻1. *Tankia tanago*: LBM1210015039.

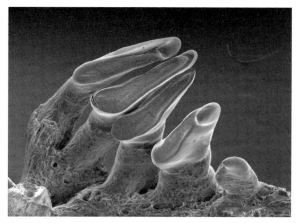

Fig. 307⁻2. *Tanakia laceolata*: LBM1210014340.

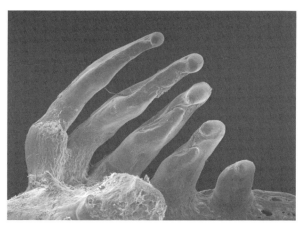

Fig. 307⁻3. *Tanakia limbata*: LBM1210015007.

Fig. 307⁻4. *Tanakia himantegus*: LBM1210015036.

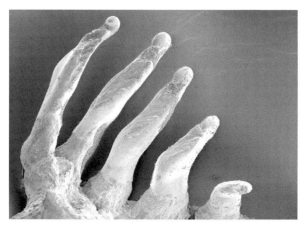

Fig. 307⁻5. *Rhodeus atremius*: LBM1210040405.

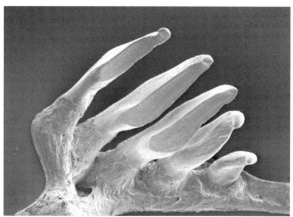

Fig. 307⁻6. *Rhodeus suigensis*: LBM1210040416.

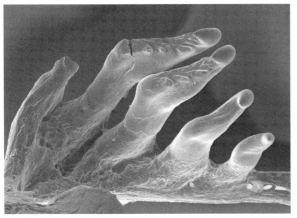

Fig. 307-7. *Rhodeus fangi*: LBM1210047475.

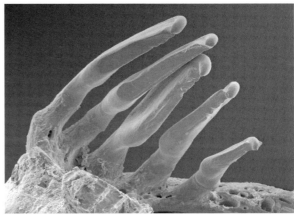

Fig. 307-8. *Rhodeus ocellatus*: LBM1210014636.

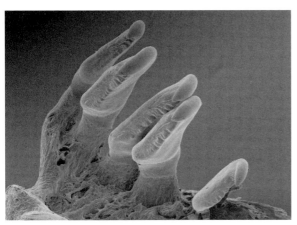

Fig. 307-9. *Rhodeus amarus*: LBM1210024448.

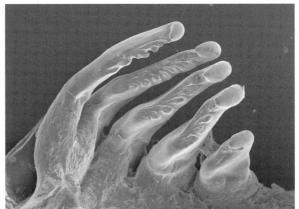

Fig. 307-10. *Rhodeus sinensis*: LBM1210014963.

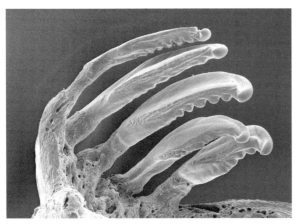

Fig. 307-11. *Acanthorhodeus chankaensis*: LBM1210014960.

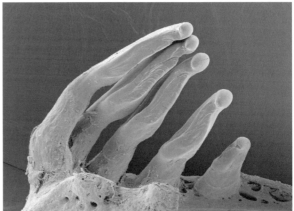

Fig. 307-12. *Acheilognathus intermedia*: LBM1210014993.

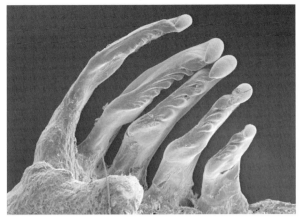

Fig. 307-13. *Acheilognathus cyanostigma*: LBM1210014969.

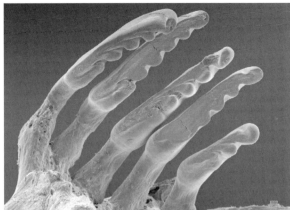

Fig. 307-14. *Acheilognathus longipinnis*: LBM1210015015.

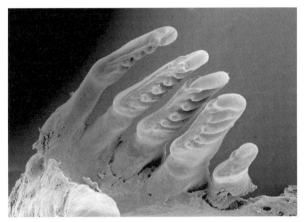

Fig. 307-15. *Acheilognathus macropterus*: LBM1210014663.

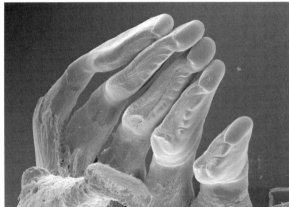

Fig. 307-16. *Acheilognathus melanogaster*: LBM1210015169.

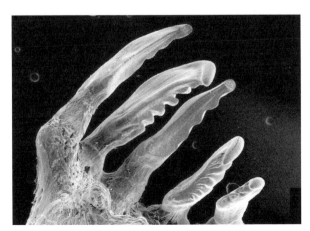

Fig. 307-17. *Acheilognathus rhombeus*: LBM1210031352.

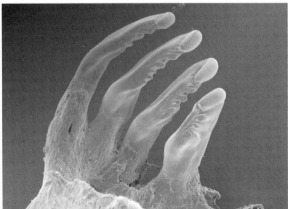

Fig. 307-18. *Acheilognathus tabira erythropterus*; LBM1210015166.

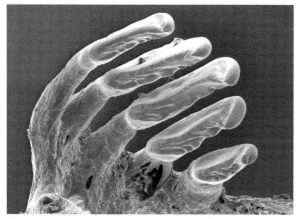

Fig. 307-19. *Acheilognathus tabira jordani*: LBM1210047474.

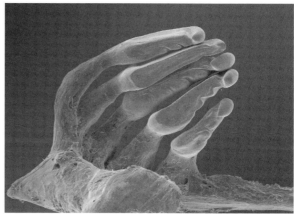

Fig. 307-20. *Acheilognathus tabira nakamurae*: LBM1210015164.

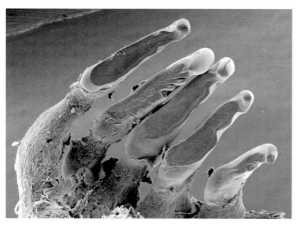

Fig. 307-21. *Acheilognathus tabira tabira*: LBM1210024465.

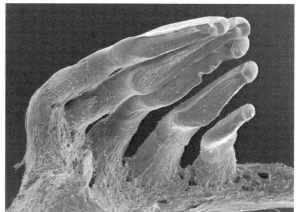

Fig. 307-22. *Acheilognathus tabira tohokuensis*: LBM1210040437.

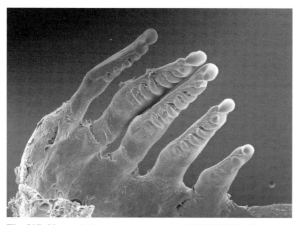

Fig. 307-23. *Acheilognathus tonkinensis*: LBM1210014660.

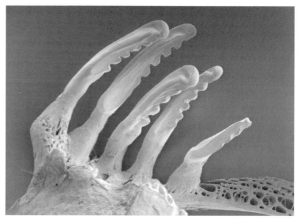

Fig. 307-24. *Acheilognathus typus*: LBM1210041993.

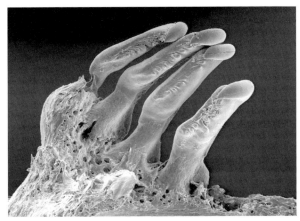

Fig. 307⁻25. *Acheilognathus yamatsutae*: LBM1210014962.

3-8. Leuciscinae

The pharyngeal dentitions of 43 species in 27 genera of Leuciscinae were examined. Their teeth are arranged in one or two rows except for three-row dentitions in *Elopichthys*, *Ochetobius*, and *Squaliobarbus*. The typical dentition is 2.4(3).5(4)-5(4).4(3).2 as in the Danioninae and Cultrinae in species with three-row dentitions, 2(1).5(4)-4(5).2(1) in species with two-row dentitions, and 4(5)-4(5) in species with one-row dentitions. In the one-row dentition of *Alburnoides*, the dental formula is 6-5, and number of major-row teeth is more than 5, as in the Xenocypridinae. Although the central teeth of most species are Type 3-0 or Type 4-0, Type 2-1 of central teeth occur in *Elopichthys*, Type 3-1 in *Ochetobius*, and Type 5-0 in *Tinca* and *Mylopharyngodon*.

Elopichthys bambusa (**Richardson, 1845**) (Fig. 308-1: IHCAS no number, BL 345.0 mm)

One specimen from Hunan, China (IHCAS no number). Dental formula: 2.4.5-4.4.2; all teeth Type 2-1. Tip of crown sharp.

Ochetobius elongatus (**Kner, 1867**) (Fig. 308-2: IHCAS73100898, BL 185.0 mm)

Four specimens from Hubei, China (IHCAS73100898). Dental formula: 2.4.4-4.4.2; central teeth Type 3-1, anterior teeth and minor-row teeth Type 2-3 or 2-1. Crowns of posterior central and minor-row teeth bent forward. Many denticles present on grinding surfaces.

Leuciscus baicalensis (**Dybowski, 1874**) (Fig. 308-3: IHCAS6290054, BL 136.0 mm)

One specimen from Xinjiang, China (IHCAS6290054). Dental formula: 2.5-5.2; central teeth Type 3-0, anterior teeth Type 3-0 or 2-3, minor-row teeth Type 2-1. Crowns of posterior central and minor-row teeth bent forward. Tooth crowns well worn with secondary grinding surfaces.

Leuciscus waleckii (**Dybowski, 1869**) (Fig. 308-4: LBM1210024352, BL 44.0 mm)

Two specimens from Siberian Russia (LBM1210024350 and 1210024352). Dental formula: 2.5-5.2; central teeth Type 3-0, anterior teeth Type 3-0 or 2-1, minor-row teeth Type 2-1. Crowns of posterior central and minor-row teeth bent forward. Denticles present on both margins of each central tooth.

Leuciscus idus (**Linnaeus, 1758**) (Fig. 308-5: LBM1210047613, BL 53.0 mm)

One specimen from Siberian Russia (LBM1210047613). Dental formula: 2.5-5.2; central teeth Type 3-0, anterior teeth Type 2-3, minor-row teeth Type 2-3 or 2-1. Crowns of posterior central and minor-row teeth bent forward. Central teeth compressed antero-posteriorly.

Phoxinus neogaeus **Cope, 1867** (Fig. 308-6: LBM1210024404, BL 44.5 mm)

One specimen from Ontario, Canada (LBM1210024404). Dental formula: 1.5-4.1; central teeth Type 3-0, anterior teeth Type 2-3 or 2-0, minor-row teeth Type 2-1. Crowns of posterior central and minor-row teeth bent forward.

Phoxinus steindachneri **Sauvage, 1883** (Fig. 308-7: LBM1210003651, BL 82.8 mm)

Five specimens from Gifu, Japan (LBM121000365-1210003653, 1210013867, and 1210013868). Dental formula: 2.(4,5)-4.2; central teeth Type 3-0, anterior teeth Type 2-3 or 2-0, minor-row teeth Type 2-3 or 2-1. Crowns of posterior central and minor-row teeth bent forward. Posterior central teeth compressed antero-posteriorly. Crowns of some teeth worn.

Phoxinus phoxinus (**Linnaeus, 1758**) (Fig. 308-8: LBM1210024408, BL 53.6 mm)

Two specimens from Siberian Russia (LBM1210024406 and 1210024408). Dental formula: (1,2).4-(4,5).(1,2); central teeth Type 3-0, anterior teeth Type 2-3 or 2-1, minor-row teeth Type 2-1. Crowns of posterior central and minor-row teeth bent forward. Anterior major- and minor-row teeth compressed antero-posteriorly.

Rhynchocypris percnurus (**Pallas, 1814**) (Fig. 308-9: LBM1210014381, BL 75.0 mm)

Four specimens from Hokkaido, Japan (LBM1210014380 to 1210014383). Dental formula: 2.4(5)-4.2; central teeth Type 3-0, anterior teeth Type 2-3 or 2-1, minor-row teeth Type 2-1. Crowns of posterior central and minor-row teeth bent forward. Denticles present on anterior magin of each central tooth.

Luxilus cornutus (**Mitchill, 1817**) (Fig. 308-10: LBM1210024354, BL 100.0 mm)

One specimen from Ontario, Canada (LBM1210024354). Dental formula: -4.2; central teeth Type 3-0, anterior teeth Type 2-3 or 2-1, minor-row teeth Type 2-1. Crowns of posterior central and minor-row teeth bent forward. Tooth crowns compressed antero-posteriorly or medio-laterally.

Lythrurus umbratilis (**Girard, 1856**) (Fig. 308-11: LBM1210024356, BL 44.0 mm)

One specimen from Ontario, Canada (LBM1210024356). Dental formula: 2.4-4.2; central teeth Type 3-0, anterior teeth and minor-row teeth Type 2-1. Crowns of posterior central and minor-row teeth bent forward. Anterior central and minor-row teeth compressed medio-laterally.

Rutilus rutilus (**Linnaeus, 1758**) (Fig. 308-12: LBM1210024450, BL 70.0 mm)

One specimen from Siberian Russia (LBM1210024450). Dental formula: 2.5-5.2; central teeth Type 3-0, anterior teeth Type 3-0 or 2-3, minor-row teeth Type 2-3 or 2-1. Crowns of posterior central and minor-row teeth bent forward. Central teeth compressed antero-posteriorly.

Pimephales promelas **Rafinesque, 1820** (Fig. 308-13: LBM1210014627, BL 69.6 mm)

Five specimens from Ontario, Canada (LBM1210024411, 1210024412, 1210014626, 1210014627, and 1210014964). Dental formula: 4-4; all teeth Type 3-0. Tooth crowns compressed antero-posteriorly.

Pimephales notatus (**Rafinesque, 1820**) (Fig. 308-14: LBM1210024410, BL 72.0 mm)

One specimen from Ontario, Canada (LBM1210024410). Dental formula: 4-4; all teeth Type 3-0. Tooth crowns compressed antero-posteriorly and well worn.

Hybognathus hankinsoni **Hubbs, 1929** (Fig. 308-15: LBM1210024339, BL 60.0 mm)

One specimen from Ontario, Canada (LBM1210024339). Dental formula: 4-4; all teeth Type 3-0. Tooth crowns compressed antero-posteriorly, well worn with secondary grinding surfaces.

Campostoma anomalum (**Rafinesque, 1820**) (Fig. 308-16: LBM1210024295, BL 75.0 mm)

One specimen from Ontario, Canada (LBM1210024295). Dental formula: 4-4; all teeth Type 3-0. Tooth crowns compressed and well worn with secondary grinding surfaces. Several denticles present on anterior margin if not eroded.

Alburnoides bipunctatus (**Bloch, 1782**) (Fig. 308-17: LBM1210026935, BL 162.1 mm)

One specimen from Europe (LBM1210026935). Dental formula: 6-5; all teeth Type 3-0. Crowns of central teeth compressed antero-posteriorly. Tooth crowns well worn with secondary grinding surfaces.

Chrosomus eos **Cope, 1861** (Fig. 308-18: LBM1210024399, BL 38.0 mm)

Two specimens from Ontario, Canada (LBM1210024397 and 1210024399). Dental formula: 5-(4,5); all teeth Type 3-0. Tooth crowns compressed antero-posteriorly. Hamulus groove present on posterior side of all tooth crowns.

Chrosomus erythrogaster (**Rafinesque, 1820**) (Fig. 308-19: LBM1210024402, BL 36.0 mm)

Two specimens from USA (LBM1210024401 and 1210024402). Dental formula: (4,5)-(4,5); all teeth Type 3-0. Tooth crowns compressed antero-posteriorly. Small hamulus groove present on posterior side of all tooth crowns. Some teeth worn.

Rhinichthys atratulus (**Hermann, 1804**) (Fig. 308-20: LBM1210024434, BL 42.5 mm)

One specimen from Ontario, Canada (LBM1210024434). Dental formula: 2.4-4.2; central teeth Type 4-0, anterior teeth Type 2-3, minor-row teeth Type 2-1. Crowns of posterior central and minor-row teeth bent forward. Some teeth worn.

Rhinichthys cataractae (**Valenciennes, 1842**) (Fig. 308-21: LBM1210024436, BL 60.0 mm)

One specimen from Ontario, Canada (LBM1210024436). Dental formula: -4.2; central teeth Type 4-0, anterior teeth Type 2-3, minor-row teeth Type 2-1. Crowns of posterior minor-row teeth bent forward.

Semotilus corporalis (**Mitchill, 1817**) (Fig. 308-22: LBM1210024458, BL 70.0 mm)

One specimen from Ontario, Canada (LBM1210024458). Dental formula: 1.4-4.2; central teeth Type 4-0 or 2-3, anterior teeth Type 2-0, minor-row teeth Type 2-1. Crowns of posterior minor-row teeth bent forward.

Ctenopharyngodon idella (**Valenciennes, 1844**) (Fig. 308-23: LBM1210013812, BL 107.0 mm)

Eight specimens from Shiga, Japan (LBM1210013612, 1210013613, 1210013650-1210013652, and 1210013655-1210013657). Dental formula: 2.(4,5)-(4,5).2; major-row teeth Type 3-0, minor-row teeth Type 2-3 or 2-1. Crowns of major-row teeth compressed antero-posteriorly. Parallel grooves present on each side of tooth crown, positional more or less alternately across longitudinal axis of crown.

Squaliobarbus curriculus (**Richardson, 1846**) (Fig. 308-24: IHCAS7747, BL 146.0 mm)

Two specimens from Shanghai, China (IHCAS7747 and LBM1210032996). Dental formula: 2.(3,4).(4,5)-4.3.2; central teeth Type 4-0 or 3-0, anterior teeth and minor-row teeth Type 2-1. Crowns of posterior central and minor-row teeth bent forward. Anterior minor-row teeth compressed medio-laterally. Hamulus groove present on posterior side of crown of central teeth. Tooth crowns well worn with secondary grinding surfaces.

Cyprinella spiloptera (**Cope, 1867**) (Fig. 308-25: LBM1210024310, BL 75.0 mm)

One specimen from Ontario, Canada (LBM1210024310). Dental formula: 1.4-5.1; central teeth Type 4-0 or 3-0, anterior teeth and minor-row teeth Type 2-0. Several denticles distributed on anterior margin of central teeth.

Margariscus margarita (**Cope, 1867**) (Fig. 308-26: LBM1210024360, BL 53.0 mm)

Two specimens from Ontario, Canada (LBM1210024358 and 1210024360). Dental formula: 2.5-4.2; major-row teeth Type 4-0, minor-row teeth Type 2-1. Crowns of posterior central teeth compressed antero-posteriorly.

Tribolodon nakamurai **Doi and Shinzawa, 2000** (Fig. 308-27: LBM1210013736, BL 302.0 mm)

Two specimens from Niigata, Japan (LBM1210013736 and 1210017712). Dental formula: 2.(4,5)-4.2; central teeth Type 4-0, anterior teeth and minor-row teeth Type 2-1. Crowns of posterior minor-row teeth bent forward.

Tribolodon brandti (**Dybowski, 1872**) (Fig. 308-28: LBM1210015893, BL 173.0 mm)

Seven specimens from Niigata (LBM1210015888, 1210015890, 1210015893, and 1210039861), Toyama (LBM1210015927 and 1210015928) and from Ibaragi, Japan (LBM1210015899). Dental forumula: 2.5-4.2; central teeth Type 4-0, anterior teeth Type 2-3 or 2-1, minor-row teeth Type 2-1. Crowns of posterior central and minor-row teeth bent forward. Some teeth worn.

Tribolodon sachalinensis (**Nikolskii, 1889**) (Fig. 308-29: LBM1210015922, BL 176.4 mm)

One specimen from Hokkaido, Japan (LBM1210015922). Dental formula: 2.5-4.2; central teeth Type 4-0, anterior teeth Type 2-3 or 2-0, minor-row teeth Type 2-1. Crowns of posterior central teeth and minor-row teeth bent forward. Some teeth worn.

Tribolodon hakonensis (**Günther, 1877**) (Fig. 308-30: LBM1210034803, BL 227.0 mm)

Seventeen specimens from Shiga, Japan (LBM1210034799-1210034815). Dental formula: 2.5-4.2; central teeth Type 4-0, anterior teeth Type 2-3 or 2-1, minor-row teeth Type 2-1. Crowns of posterior central and minor-row teeth bent forward, those of major-row teeth compressed antero-posteriorly.

Clinostomus elongatus (**Kirtland, 1840**) (Fig. 308-31: LBM1210024302, BL 45.0 mm)

One specimen from Ontario, Canada (LBM1210024302). Dental formula: 2.(4,5)-4.2; central teeth Type 4-0, anterior teeth Type 2-3 or 2-1, minor-row teeth Type 2-1. Crowns of posterior central and minor-row teeth bent forward. Tooth crowns compressed antero-posteriorly or medio-laterally.

Couesius plumbeus (**Agassiz, 1850**) (Fig. 308-32: LBM1210024306, BL 49.0 mm)

Two specimens from Ontario, Canada (LBM1210024304 and 1210024306). Dental formula:

2.4-(1,2).4; central teeth Type 4-0, anterior teeth and minor-row teeth Type 2-1. Crowns of posterior central and minor-row teeth bent forward. Crowns of major row teeth compressed antero-posteriorly or medio-laterally.

Scardinius erythrophthalmus **(Linnaeus, 1758)** (Fig, 308-33: LBM1210043997, BL 87.3 mm)

Three specimens from Shiga, Japan (LBM1210043995, 1210043997, and 1210043999). Dental formula: 2(3).4(5)-5(4).3(2); central teeth Type 4-0, anterior teeth Type 4-0 or 2-1, minor-row teeth Type 2-1. Crowns of posterior central and minor-row teeth bent forward. Anterior margin of major-row teeth bearing large denticles, appearing serrate if not eroded. Crowns of central teeth compressed antero-posteriorly. Some teeth worn.

Notropis atherinoides **Rafinesque, 1818** (Fig. 308-34: LBM1210024373, BL 75.0 mm)

One specimen from Ontario, Canada (LBM1210024373). Dental formula: 2.4-4.2; central teeth Type 4-0, anterior teeth Type 2-3, minor-row teeth Type 2-1. Tooth crowns compressed antero-posteriorly or medio-laterally. Denticles present on both margins of each central tooth.

Notropis heterodon **(Cope, 1865)** (Fig. 308-35: LBM1210024375, BL 37.0 mm)

One specimen from Ontario, Canada (LBM1210024375). Dental formula: 1.4-4.1; central teeth Type 4-0, anterior teeth Type 2-3, minor-row teeth Type 2-1. Crowns of posterior central and minor-row teeth bent forward. Crowns of major-row teeth compressed antero-posteriorly or medio-laterally, with several denticles on anterior margin.

Notropis rubellus **(Agassiz, 1850)** (Fig. 308-36: LBM1210024379, BL 32.5 mm)

One specimen from Ontario, Canada (LBM1210024379). Dental formula: 2.5-4.2; central teeth Type 4-0, anterior teeth Type 2-3 or 2-1, minor-row teeth Type 2-1. Crowns of posterior central and minor-row teeth bent forward. Crowns of central teeth compressed antero-posteriorl.

Notropis volucellus **(Cope, 1865)** (Fig. 308-37: LBM1210024383, BL 50.0 mm)

One specimen from Ontario, Canada (LBM1210024383). Dental formula: 4-4; central teeth Type 4-0, anterior teeth Type 2-3. Denticles present on anterior margins of central teeth.

Notropis heterolepis **Eigenmann and Eigenmann, 1893** (Fig. 308-38: LBM1210024377, BL 54.0 mm)

One specimen from Ontario, Canada (LBM1210024377). Dental formula: 4-4; all teeth Type 4-0. Denticles present on both margins of each tooth.

Nocomis biguttatus **(Kirtland, 1840)** (Fig. 308-39: LBM1210024366, BL 95.0 mm)

One specimen from Ontario, Canada (LBM1210024366). Dental formula: 4-4; central teeth Type 4-0, anterior teeth Type 2-0. Crowns of posterior central teeth bent forward.

Nocomis micropogon **(Cope, 1865)** (Fig. 308-40: LBM1210024367, BL 100.0 mm)

One specimen from Ontario, Canada (LBM1210024367). Dental formula: 4-4; central teeth Type 4-0, anterior teeth Type 2-0. Crowns of posterior central teeth bent forward. Some teeth worn.

Notemigonus crysoleucas **(Mitchill, 1814)** (Fig. 308-41: LBM1210024371, BL 63.0 mm)

One specimen from Ontario, Canada (LBM1210024371). Dental formula: -5; central teeth Type 4-0, anterior teeth Type 4-0 or 2-3. Tooth crowns of teeth A2 (An2) to A5 (Po1) compressed antero-posteriorly. Denticles present on anterior margin and tiny denticles on grinding surface of teeth A2 (An2) to A5 (Po1).

Tinca tinca **(Linnaeus, 1758)** (Fig. 308-42: LBM1210013973, BL 154.0 mm)

Eight specimens from Italy (LMB1210013973, 1210015053, 1210015155, 1210021930, 1210021931, 1210036781-1210036783). Dental formula: 5(4)-5(4); central teeth Type 5-0 or Type 4-0, anterior teeth are Type 5-0. Groove along anterior margin of teeth An2 (A2) to Po1 (A5). Denticles present on posterior margin of each central tooth. Some teeth well worn.

Mylopharyngodon piceus **(Richardson, 1846)** (Fig. 308-43: LBM1210013668, BL 820.0 mm)

Eight specimens from Saitama, Japan (LBM1210013645, 1210013648, 1210013649, 1210013668, 1210013669, and 1210013678-

1210013680). Dental formula: 4–5; all teeth Type 5–0 or 2–0. Teeth massive and depressed. Very shallow transverse groove present along anterior margin on grinding surface of central teeth. All teeth massive.

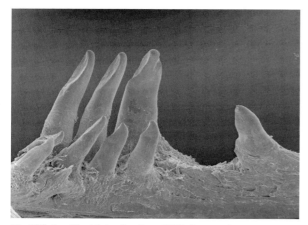

Fig. 308-1. *Elopichthys bambusa*: IHCAS no number.

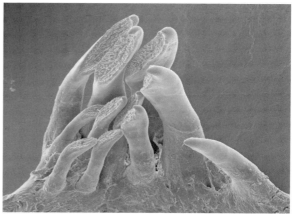

Fig. 308-2. *Ochetobius elongates*: IHCAS73100898.

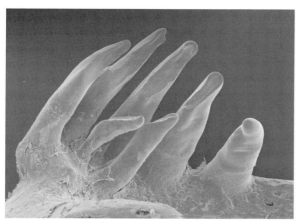

Fig. 308-3. *Leuciscus baicalensis*: IHCAS6290054.

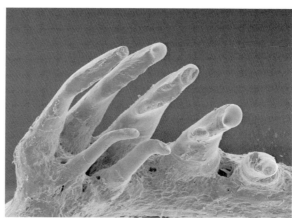

Fig. 308-4. *Leuciscus waleckii*: LBM1210024352.

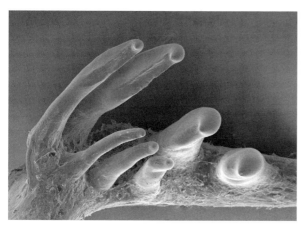

Fig. 308-5. *Leuciscus idus*: LBM1210047613.

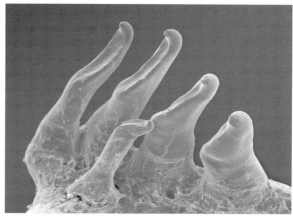

Fig. 308-6. *Phoxinus neogaeus*: LBM1210024404.

3. Descriptions of the pharyngeal dentition of cyprinid subfamilies   105

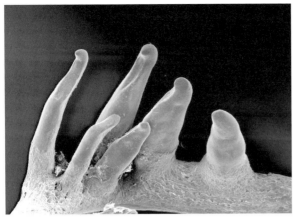

Fig. 308-7. *Phoxinus steindachneri*: LBM1210003651.

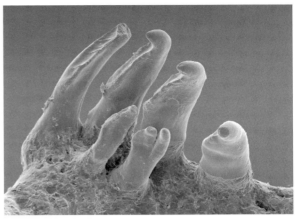

Fig. 308-8. *Phoxinus phoxinus*: LBM1210024408.

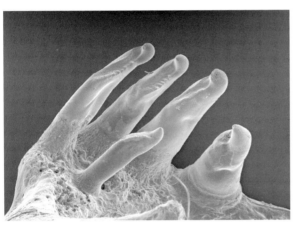

Fig. 308-9. *Rhynchocypris percnurus*: LBM1210014381.

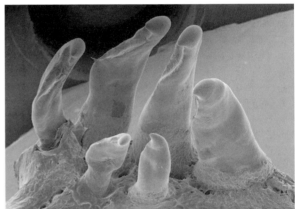

Fig. 308-10. *Luxilus cornutus*: LBM1210024354.

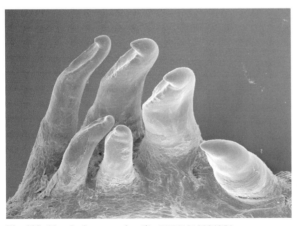

Fig. 308-11. *Lythrurus umbratilis*: LBM1210024356.

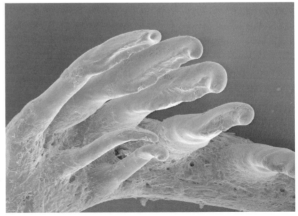

Fig. 308-12. *Rutilus rutilus*: LBM1210024450.

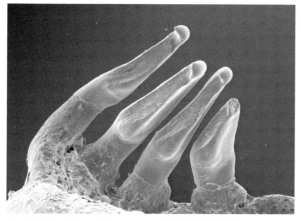

Fig. 308-13. *Pimephales promelas*: LBM1210014627.

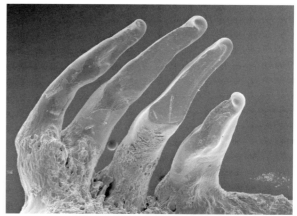

Fig. 308-14. *Pimephales notatus*: LBM1210024410.

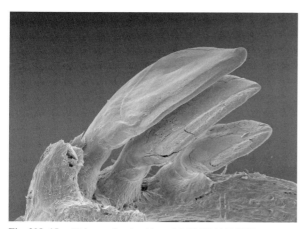

Fig. 308-15. *Hybognathus hankinsoni*: LBM1210024339.

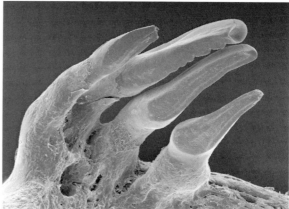

Fig. 308-16. *Campostoma anomalum*: LBM1210024295.

Fig. 308-17. *Alburnoides bipunctatus*: LBM1210026935.

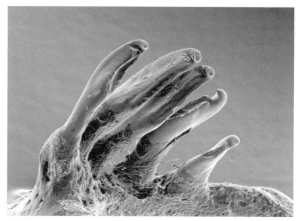

Fig. 308-18. *Chrosomus eos*: LBM1210024399.

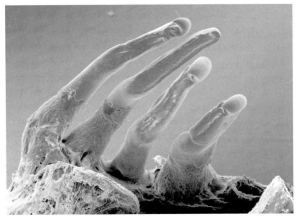

Fig. 308⁻19. *Chrosomus erythrogaster*: LBM1210024402.

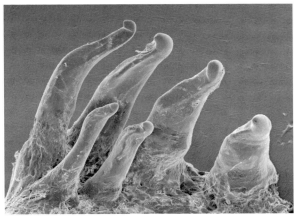

Fig. 308⁻20. *Rhinichthys atratulus*: LBM1210024434.

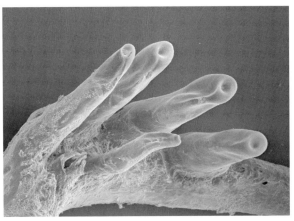

Fig. 308⁻21. *Rhinichthys cataractae*: LBM1210024436.

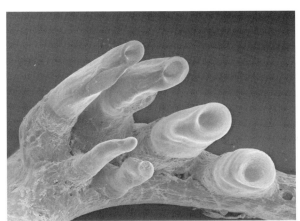

Fig. 308⁻22. *Semotilus corporalis*: LBM1210024458.

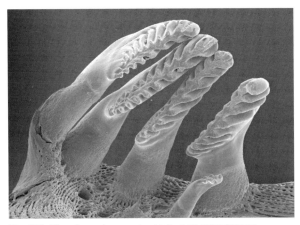

Fig. 308⁻23. *Ctenopharyngodon idella*: LBM1210013812.

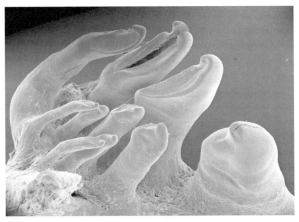

Fig. 308⁻24. *Squaliobarbus curriculus*: IHCAS7747.

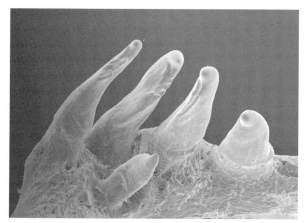

Fig. 308−25. *Cyprinella spiloptera*: LBM1210024310.

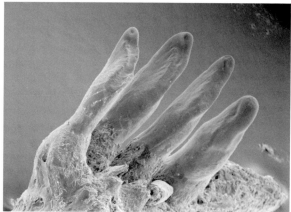

Fig. 308−26. *Margariscus margarita*: LBM1210024360.

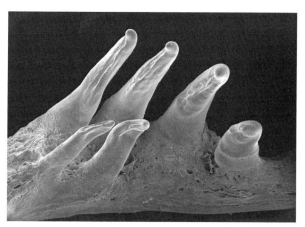

Fig. 308−27. *Tribolodon nakamurai*: LBM1210013736.

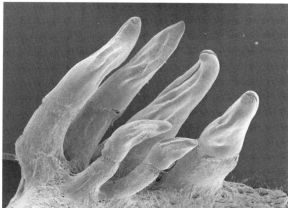

Fig. 308−28. *Tribolodon brandti*: LBM1210015893.

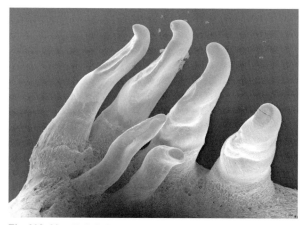

Fig. 308−29. *Tribolodon sachalinensis*: LBM1210015922.

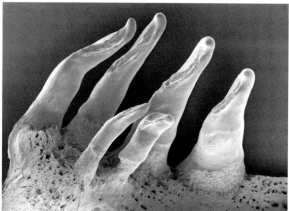

Fig. 308−30. *Tribolodon hakonensis*: LBM1210034803.

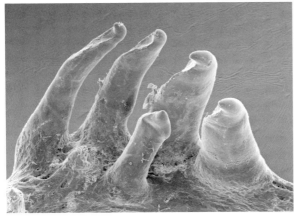

Fig. 308-31. *Clinostomus elongatus*: LBM1210024302.

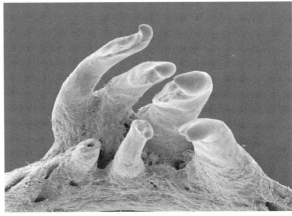

Fig. 308-32. *Couesius plumbeus*: LBM1210024306.

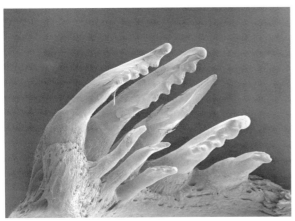

Fig. 308-33. *Scardinius erythrophthalmus*: LBM1210043997.

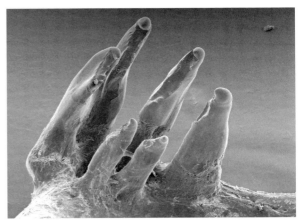

Fig. 308-34. *Notropis atherinoides*: LBM1210024373.

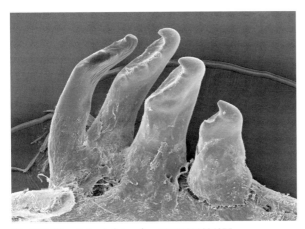

Fig. 308-35. *Notropis heterodon*: LBM1210024375.

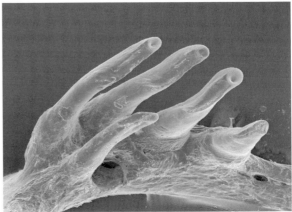

Fig. 308-36. *Notropis rubellus*: LBM1210024379.

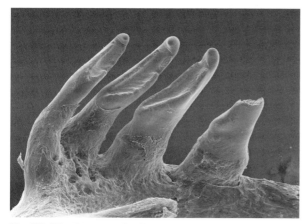

Fig. 308⁻37. *Notropis volucellus*: LBM1210024383.

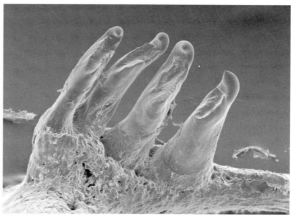

Fig. 308⁻38. *Notropis heterolepis*: LBM1210024377.

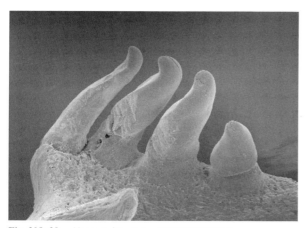

Fig. 308⁻39. *Nocomis biguttatus*: LBM1210024366.

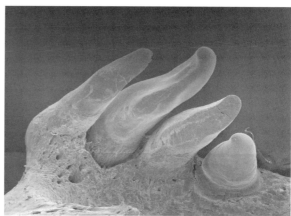

Fig. 308⁻40. *Nocomis micropogon*: LBM1210024367.

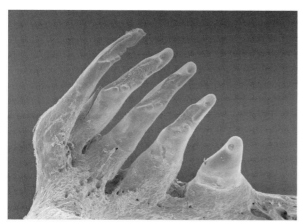

Fig. 308⁻41. *Notemigonus crysoleucas*: LBM1210024371.

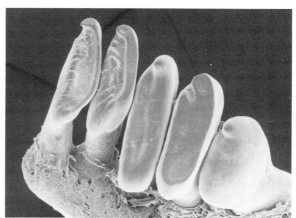

Fig. 308⁻42. *Tinca tinca*: LBM1210013973.

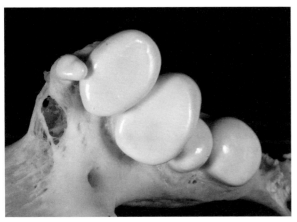

Fig. 308-43. *Mylopharyngodon piceus*: LBM1210013668.

3-9. Labeoninae

The pharyngeal dentitions of 33 species in 12 genera of Labeoninae were examined. Their teeth are always arranged in three rows; the typical dental formula is 2.4(3).5-5.4(3).2 as in danionine and cultrine dentitions; the central teeth are generally Type 4-1. Most labeonine teeth are well worn, and secondary grinding surfaces are frequently formed. Hamulus grooves are usually present on major-row teeth, but it cannot be observed in worn teeth in which a secondary grinding surface has been formed.

Barbichthys laevis (**Valenciennes, 1842**) (Fig. 309-1: LBM1210014062, BL 40.5 mm)

One specimen from Thailand (LBM1210014062). Dental formula: 2.4.5-5.4.2; all teeth Type 4-1. Crowns of posterior central and minor-row teeth bent forward and compressed antero-posteriorly. Teeth well worn with secondary grinding surfaces.

Catla catla (**Hamilton, 1822**) (Fig. 309-2: LBM1210026892, BL37.2 mm)

One specimen from Bangladesh (LBM1210026892). Dental formula: 2.4.5-5.4.2; major-row teeth Type 4-1, minor-row teeth Type 4-1 or 2-0. Crowns of posterior central and minor-row teeth bent forward and compressed antero-posteriorly. Tiny hamulus groove present on posterior side of crown of major-row teeth. Minor-row teeth slender. Crowns of some teeth well worn.

Garra cambodgiensis (**Tirant, 1883**) (Fig. 309-3: LBM1210026900, BL 54.4 mm)

Two specimens from Cambodia (LBM1210026896 and 1210026900). Dental formula: 2.4.5-5.4.2; central teeth Type 4-1, anterior teeth and minor-row teeth Type 4-1 or 2-1. Crowns of posterior central and minor-row teeth bent forward and compressed antero-posteriorly. Minor-row teeth slender. Crowns of some teeth worn.

Garra gotyla gotyla (**Gray, 1830**) (Fig. 309-4: LBM1210026888, BL 106.7 mm)

One specimen from Nepal (LBM1210026888). Dental formula: 2.4.5-5.4.2; central teeth and minor-row teeth Type 4-1 or 2-1, anterior teeth Type 2-1. Tooth crowns bent forward. Shallow grooves along anterior or lateral margins of grinding surfaces. Minor-row teeth slender. Crowns of some teeth worn.

Garra orientalis **Nichols, 1925** (Fig. 309-5: LBM1210024317, BL 100.0 mm)

Three specimens from China (LBM1210024313, 1210024315, and 1210024317). Dental formula: 2.4.5-5.3.2; central teeth and minor-row teeth Type 4-1 or 2-1 anterior teeth Type 2-1. Crowns of posterior central and minor-row teeth bent forward and compressed antero-posteriorly. Minor-row teeth slender. Crowns of some teeth worn.

Labeo chrysophekadion (**Bleeker, 1849**) (Fig. 309-6: LBM1210015735, BL 29.5 mm)

Three specimens from Thailand (LBM1210013858, 1210013995, and 1210014038). Dental formula: 2.4.5-5.4.2; central teeth Type 4-1, anterior teeth and minor-row teeth Type 4-1 or 2-1. Crowns of posterior central and minor-row teeth bent forward and compressed antero-posteriorly. Minor-row teeth slender. Crowns of some teeth worn.

Labeo boga (**Hamilton, 1822**) (Fig. 309-7: LBM1210015738. BL 97.5 mm)

One specimen from India (LBM1210015738). Dental formula: 2.3.5-5.4.2; central teeth Type 4-1, anterior teeth Type 4-1 or 2-1, minor-row teeth Type 4-1 or 2-3. Crowns of posterior central and minor-row teeth bent forward and compressed antero-posteriorly. Teeth well worn with secondary grinding surfaces.

Labeo cylindricus **Peters, 1852** (Fig. 309-8: LBM1210014043, BL 131.0 mm)

One specimen from Malawi (LBM1210013992). Dental formula: -5.4.2; central teeth and minor-row teeth Type 4-1, anterior teeth Type 4-1 or 2-1. Crowns of posterior central and minor-row teeth bent forward and compressed antero-posteriorly. Teeth well worn with secondary grinding surfaces.

Labeo erythropterus **Valenciennes, 1842** (Fig. 309-9: LBM1210013847, BL 114.5 mm)

Two specimens from Thailand (LBM1210013847 and 1210013855). Dental formula: 2.4.5-5.4.2; major-row teeth Type 4-1, minor-row teeth Type 4-1 or 2-3. Crowns of posterior central and minor-row teeth bent forward and compressed antero-posteriorly. Teeth well worn with secondary grinding surfaces.

Labeo rohita (**Hamilton, 1822**) (Fig. 309-10: LBM1210014049, BL 115.0 mm)

One specimen from India (LBM1210014049). Dental formula: -5.4.2; major-row teeth Type 4-1, minor-row teeth Type 4-1 or 2-3. Crowns of posterior central and minor-row teeth bent forward and compressed antero-posteriorly. Teeth well worn with secondary grinding surfaces.

Labeo rosae **Steindachner, 1894** (Fig. 309-11: LBM1210024340, BL 108.6 mm)

One specimen from South Africa (LBM1210024340). Dental formula: 2.4.5-5.4.2; central teeth and minor-row teeth Type 4-1, anterior teeth Type 4-1 or 2-1. Crowns of posterior central and minor-row teeth bent forward and compressed antero-posteriorly. Teeth well worn with secondary grinding surfaces.

Cirrhinus mrigala (**Hamilton, 1822**) (Fig. 309-12: LBM1210014083, BL 73.0 mm)

One specimen from Nepal (LBM1210014083). Dental formula: 2.4.5-5.4.2; all teeth Type 4-1. Crowns of posterior central and minor-row teeth bent forward and compressed antero-posteriorly. Teeth well worn with secondary grinding surfaces.

Cirrhinus microlepis **Sauvage, 1878** (Fig. 309-13: LBM1210047493, BL 148.1 mm)

One specimen from Cambodia (LBM1210047493). Dental formula: 2.4.4-5.3.1; all teeth Type 4-1. Crowns of posterior central and minor-row teeth bent forward. Tooth crowns well worn with secondary grinding surfaces.

Cirrhinus molitorella (**Valenciennes, 1844**) (Fig. 309-14: LBM1210014031, BL 204.0 mm)

Two specimens from Hubei, China (LBM1210014031 and 1210026886). Dental formula: 2.4.5-5.4.2; all teeth Type 4-1. Crowns of posterior central and minor-row teeth bent forward. Teeth intensely worn with secondary grinding surfaces.

Lobocheilos melanotaenia (**Fowler, 1935**) (Fig. 309-15: LBM1210047497, BL 81.2 mm)

Two specimens from Cambodia (LBM1210047497 and 1210047508). Dental formula: 2.4.5-5.4.2; all teeth Type 4-1. Crowns of posterior central and minor-row teeth bent forward and compressed antero-posteriorly. Small hamulus groove present on posterior side of crown of central and minor-row teeth. Teeth well worn with secondary grinding surfaces.

Lobocheilos rhabdoura (**Fowler, 1934**) (Fig. 309-16: LBM1210026898, BL 52.0 mm)

One specimen from Cambodia (LBM1210026898). Dental formula: 2.3.5-5.3.2; all teeth Type 4-1. Crowns of posterior central and minor-row teeth bent forward and compressed antero-posteriorly. Teeth well worn with secondary grinding surfaces. Hamulus groove present on posterior side of crown.

Henicorhynchus siamensis (**Sauvage, 1881**) (Fig. 309-17: LBM1210026942, BL 94.9 mm)

One specimen from Cambodia (LBM1210026942). Dental formula: 2.4.5-5.4.2; central teeth and minor-row teeth Type 4-1, anterior teeth Type 4-1 or 2-1. Crowns of posterior central and minor-row teeth bent forward and compressed antero-posteriorly. Teeth well worn with secondary grinding surfaces.

Epalzeorhynchos munense (**Smith, 1934**) (Fig. 309-18: LBM1210047611, BL 53.5 mm)

One specimen from Cambodia (LBM1210047611). Dental formula: 2.4.5-5.4.2; all teeth Type 4-1. Crowns of posterior central and minor-row teeth bent forward and compressed antero-posteriorly. Hamulus groove present on posterior side of crown of major-row teeth. Crowns of some teeth worn.

Epalzeorhynchos bicolor (**Smith, 1931**) (309-19: LBM1210015733, BL 35.5 mm)

Two specimens from Thailand (LBM1210015733 and 1210026926). Dental formula: 2.4.5-5.4.2; major-row teeth Type 4-1, minor-row teeth Type 4-1 or 2-1. Crowns of posterior central and minor-row teeth bent forward and compressed antero-posteriorly. Hamulus groove present on posterior side of crown of major-row teeth. Minor-row teeth slender. Crowns of some teeth worn.

Epalzeorhynchos frenatus (**Fowler, 1934**) (Fig. 309-20: LBM1210015731, BL 37.4 mm)

One specimen from Thailand (LBM1210015731). Dental formula: 2.4.5-5.4.2; all teeth Type 4-1. Crowns of posterior central and minor-row teeth bent forward and compressed antero-posteriorly. Hamulus groove present on posterior side of crown of major-row teeth. Crowns of some teeth worn.

Rectoris luxiensis **Wu & Yao, 1977** (Fig. 309-21: IHCAS no number, BL 72.0 mm)

Three specimens from Hubei, China (IHCAS8190849, 8190476, and no number). Dental formula: 2.4.5-5.4.2; all teeth Type 4-1. Crowns of posterior central and minor-row teeth bent forward and compressed antero-posteriorly. Teeth well worn with secondary grinding surfaces.

Crossocheilus oblongus **Kuhl and Van Hasselt, 1823** (Fig. 309-22: LBM1210013992, BL 69.0 mm)

Two specimens from Thailand (LBM1210013992) and Malaysia (LBM1210028158). Dental formula: 2.4.5-5.4.2; all teeth Type 4-1. Significant hamulus groove present on posterior side of crown of major-row teeth. Crowns of some teeth worn.

Osteochilus lini **Fowler, 1935** (Fig. 309-23: LBM1210047505, BL 79.9 mm)

Three specimens from Cambodia (LBM1210047485, 1210047505, and 1210047507). Dental formula: 2.4.5-5.4.2; all teeth Type 4-1. Crowns of posterior central and minor-row teeth bent forward and compressed antero-posteriorly. Significant hamulus groove present on posterior side of crown of major-row teeth. Teeth well worn with secondary grinding surfaces.

Osteochilus melanopleurus **(Bleeker, 1852)** (Fig. 309-24: LBM1210014033, BL 88.0 mm)

Two specimens from Thailand (LBM1210014032 and 1210014033). Dental formula: 2.(3,4).5-5.4.2; all teeth Type 4-1. Crowns of posterior central and minor-row teeth bent forward and compressed antero-posteriorly. Significant hamulus groove present on posterior side of crown of major-row teeth. Teeth well worn with secondary grinding surfaces.

Osteochilus microcephalus **(Valenciennes, 1842)** (Fig. 309-25: LBM1210033016, BL 104.8 mm)

Two specimens from Malaysia (LBM1210024391) and Cambodia (LBM1210033016). Dental formula: 2.4.5-5.4.2; all teeth Type 4-1. Crowns of posterior central and minor-row teeth bent forward and compressed antero-posteriorly. Significant hamulus groove present on posterior side of crown of major-row teeth. Teeth well worn.

Osteochilus salsburyi **Nichols & Pope, 1927** (Fig. 309-26: IHCAS824327, BL 142.0 mm)

Three specimens from Hainan, China (IHCAS602707, 7654211, and 8243271). Dental formula: 2.4.5-5.4.2; all teeth Type 4-1. Crowns of posterior central and minor-row teeth bent forward and compressed antero-posteriorly. Significant hamulus groove present on posterior side of crown of major-row teeth. Teeth well worn with secondary grinding surfaces.

Osteochilus spilurus **(Bleeker, 1851)** (Fig. 309-27: LBM1210024389, BL 50.5 mm)

Three specimens from Malaysia (LBM1210024385, 1210024387, and 1210024389). Dental formula: 2.4.5-5.4.2; all teeth Type 4-1. Crowns of posterior central and minor-row teeth bent forward and compressed antero-posteriorly. Significant hamulus groove present on posterior side of crown of major-row teeth. Teeth well worn with secondary grinding surfaces.

Osteochilus vittatus **(Valenciennes, 1842)** (Fig. 309-28: LBM1210014053, BL 107. 0 mm)

Three specimens from Thailand (LBM1210014034 1210014035, and 1210014053). Dental formula: 2.4.5-5.4.2; all teeth Type 4-1. Crowns of posterior central and minor-row teeth bent forward and compressed antero-posteriorly. Significant hamulus groove present on posterior side of crown of major-row teeth. Teeth well worn with secondary grinding surfaces.

Osteochilus waandersii **(Bleeker, 1853)** (Fig. 309-29: LBM1210047499, BL 102.5 mm)

Two specimens from Cambodia (LBM1210047499 and 1210047501). Dental formula: 2.4.5-5.4.2; all teeth Type 4-1. Crowns of posterior central and minor-row teeth bent forward and compressed antero-posteriorly. Significant hamulus groove present on posterior side of crown of major-row teeth. Teeth well worn with secondary grinding surfaces.

Labiobarbus festivus **(Heckel, 1843)** (Fig. 309-30: LBM1210024348, BL 111.0 mm)

Three specimens from Malaysia (LBM1210024344, 1210024346, and 1210024348). Dental formula: 2.4.5-5.4.2; all teeth Type 4-1. Crowns of posterior central and minor-row teeth bent forward and compressed antero-posteriorly. Significant hamulus groove present on posterior side of crown of major-row teeth. Teeth well worn with secondary grinding surfaces.

***Labiobarbus leptocheilus* (Valenciennes, 1842)** (Fig. 309⁻31: LBM1210014059, BL 128.0 mm)

One specimen from Thailand (LBM1210014059). Dental formula: 2.4.5⁻5.4.2; all teeth Type 4⁻1. Crowns of posterior central and minor-row teeth bent forward and compressed antero-posteriorly. Significant hamulus groove present on posterior side of crown of major-row teeth. Teeth well worn with secondary grinding surfaces.

***Labiobarbus lineatus* (Sauvage, 1878)** (Fig. 309⁻32: LBM1210047503, BL 85.7 mm)

One specimen from Cambodia (LBM1210047503). Dental formula: 2.4.5⁻5.4.2; all teeth Type 4⁻1. Crowns of posterior central and minor-row teeth bent forward and compressed antero-posteriorly. Significant hamulus groove present on posterior side of crown of major-row teeth. Teeth well worn with secondary grinding surfaces.

***Labiobarbus siamensis* (Sauvage, 1881)** (Fig. 309⁻33: LBM1210047491, BL 99.3 mm)

One specimen from Cambodia (LBM1210047491). Dental formula: 2.4.5⁻5.4.2; all Type 4⁻1. Crowns of posterior central and minor-row teeth bent forward and compressed antero-posteriorly. Significant hamulus groove present on posterior side of crown of major-row teeth. Teeth well worn with secondary grinding surfaces.

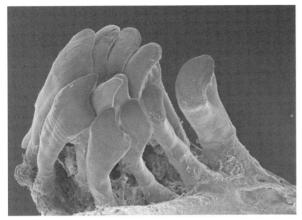

Fig. 309-1. *Barbichthys laevis*: LBM1210014062.

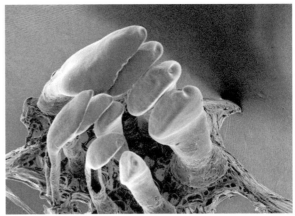

Fig. 309-2. *Catla catla*: LBM1210026892.

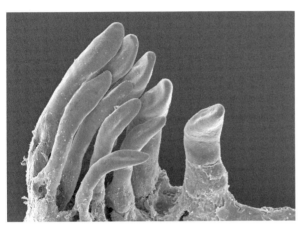

Fig. 309-3. *Garra cambodgiensis*: LBM1210026900.

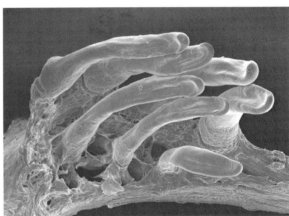

Fig. 309-4. *Garra gotyla gotyla*: LBM1210026888.

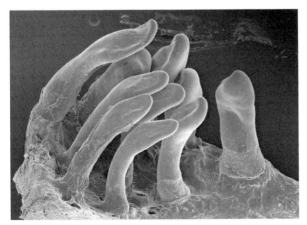

Fig. 309-5. *Garra orientalis*: LBM1210024317.

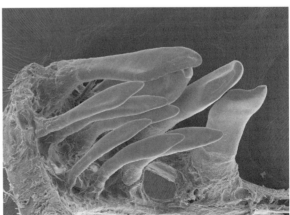

Fig. 309-6. *Labeo chrysophekadion*: LBM1210015735.

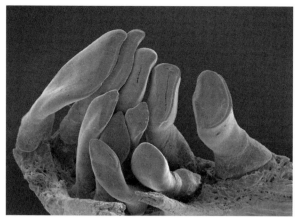

Fig. 309-7. *Labeo boga*: LBM1210015738.

Fig. 309-8. *Labeo cylindricus*: LBM1210014043.

Fig. 309-9. *Labeo erythropterus*: LBM1210013847.

Fig. 309-10. *Labeo rohita*: LBM1210014049.

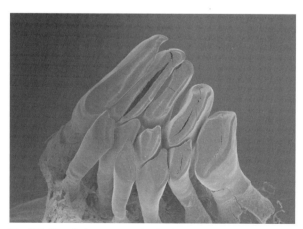

Fig. 309-11. *Labeo rosae*: LBM1210024340.

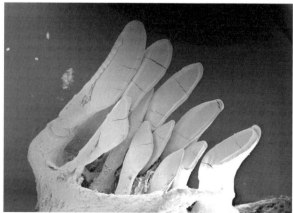

Fig. 309-12. *Cirrhinus mrigala*: LBM1210014083.

Fig. 309-13. *Cirrhinus microlepis*: LBM1210047493.

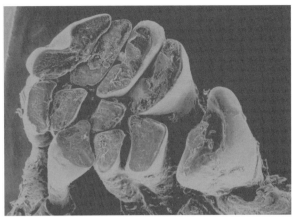

Fig. 309-14. *Cirrhinus molitorella*: LBM1210014031.

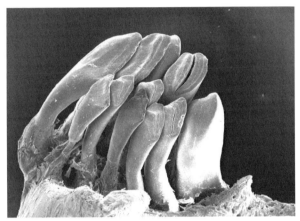

Fig. 309-15. *Lobocheilos melanotaenia*: LBM1210047497.

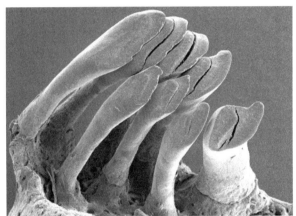

Fig. 309-16. *Lobocheilos rhabdoura*: LBM1210026898.

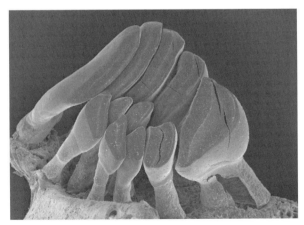

Fig. 309-17. *Henicorhynchus siamensis*: LBM1210026942.

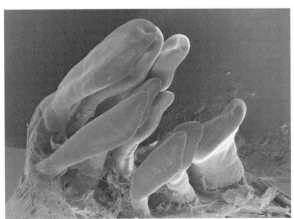

Fig. 309-18. *Epalzeorhynchos munense*: LBM1210047611.

3. Descriptions of the pharyngeal dentition of cyprinid subfamilies

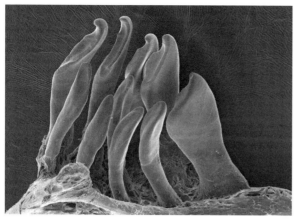

Fig. 309-19. *Epalzeorhynchos bicolor*: LBM1210015733.

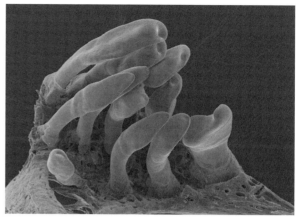

Fig. 309-20. *Epalzeorhynchos frenatus*: LBM1210015731.

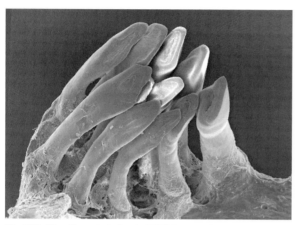

Fig. 309-21. *Rectoris luxiensis*: IHCAS no number.

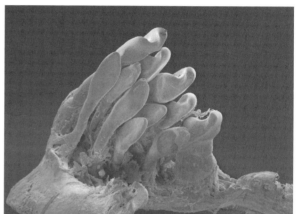

Fig. 309-22. *Crossocheilus oblongus*: LBM1210013992.

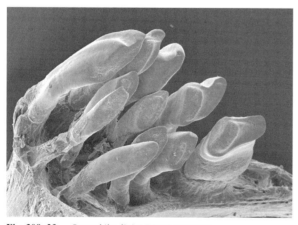

Fig. 309-23. *Osteochilus lini*: LBM1210047505.

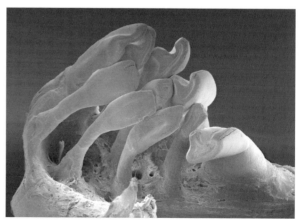

Fig. 309-24. *Osteochilus melanopleurus*: LBM1210014033.

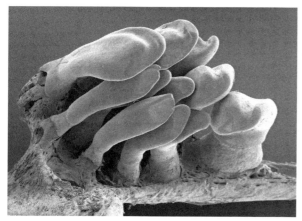
Fig. 309-25. *Osteochilus microcephalus*: LBM1210033016.

Fig. 309-26. *Osteochilus salsburyi*: IHCAS824327.

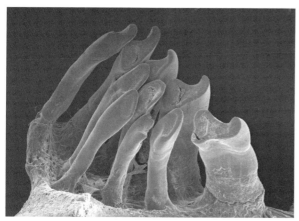
Fig. 309-27. *Osteochilus spilurus*: LBM1210024389.

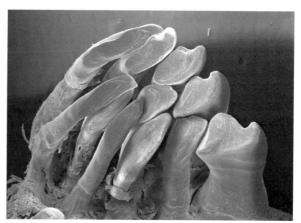
Fig. 309-28. *Osteochilus vittatus*: LBM1210014053.

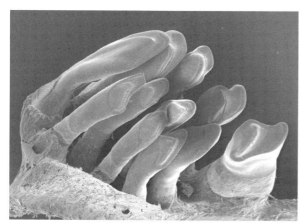
Fig. 309-29. *Osteochilus waandersii*: LBM1210047499.

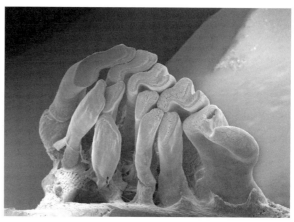
Fig. 309-30. *Labiobarbus festivus*: LBM1210024348.

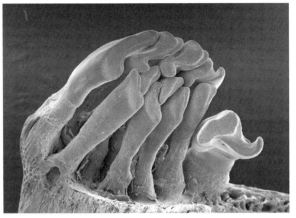

Fig. 309-31. *Labiobarbus leptocheilus*: LBM1210014059.

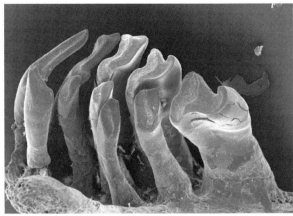

Fig. 309-32. *Labiobarbus lineatus*: LBM1210047503.

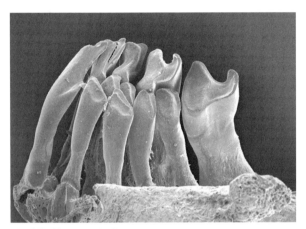

Fig. 309-33. *Labiobarbus siamensis*: LBM1210047491.

3-10. Gobiobotinae

The pharyngeal dentition of nine species in two genera in Gobiobotinae were examined. Their teeth are arranged in two rows. The typical dental formula is 3(2).5(4)-5(4).3(2). Variation in tooth number takes place rarely: four teeth in the major row and two teeth in the minor row. The central teeth are Type 4-1, and the anterior teeth and minor-row teeth are Type 2-1 in nearly all species.

Gobiobotia abbreviata **Fang and Wang, 1931** (Fig. 310-1: IHCAS12208010, BL 78.6 mm)

Two specimens from Sichuan, China (IHCAS12208010). Dental formula: 3.5-5.3; central teeth Type 4-1, anterior teeth Type 2-0, minor-row teeth Type 2-1. Crowns of teeth at posterior positions bent forward. Some teeth worn.

Gobiobotia brevibarba **Mori, 1935** (Fig. 310-2: LBM1210017560, BL 80.0 mm)

One specimen from Korea (LBM1210017560). Dental formula: 3.4-4.3; central teeth Type 4-1, anterior teeth and minor-row teeth Type 2-1. Crowns of teeth at posterior positions bent forward.

Gobiobotia filifer **(Garman, 1912)** (Fig. 310-3: IHCAS no number, BL 91.3 mm)

Three specimens from Sichuan, China (IHCAS no numbers). Dental formula: 3(2).5-5.3; central teeth Type 4-1, anterior teeth and minor-row teeth Type 2-1. Crowns of teeth at posterior positions bent forward.

Gobiobotia guilingensis **Chen, 1989** (Fig. 310-4: IHCAS12208060, BL 74.5 mm)

Two specimens from Guizhou, China (IHCAS12208060). Dental formula: 2.5-5.2; central teeth Type 4-1, anterior teeth and minor-row teeth Type 2-1. Crowns of teeth at posterior positions bent forward.

Gobiobotia kolleri **Bănărescu and Nalbant, 1966** (Fig. 310-5: IHCAS12208019, BL 68.8 mm)

Two specimens from Fujian, China (IHCAS12208019). Dental formula: 3.5-5.3; central teeth Type 4-1, anterior teeth and minor-row teeth Type 2-1. Crowns of teeth at posterior positions bent forward.

Gobiobotia meridionalis **Chen and Cao, 1977** (Fig. 310-6: IHCAS12208018, BL 62.8 mm)

Two specimens from Guangdong, China (IHCAS12208018). Dental formula: 3.5-5.3; central teeth Type 4-1, anterior teeth and minor-row teeth Type 2-1 or 2-0. Crowns of teeth at posterior positions bent forward.

Gobiobotia pappenheimi **Kreyenberg, 1911** (Fig. 310-7: IHCAS12208046, BL 52.8 mm)

Two specimen from Henan, China (IHCAS12208046). Dental formula: 3.5-5.3; central teeth Type 4-1, anterior teeth and minor-row teeth Type 2-1. Crowns of teeth at posterior positions bent forward.

Gobiobotia tungi **Fang, 1933** (Fig. 310-8: IHCAS12208006, BL 91.2 mm)

Two specimens from Zhejiang, China (IHCAS12208006). Dental formula: 3.5-5.3; central teeth Type 4-1, anterior teeth and minor-row teeth Type 2-1 or 2-0. Crowns of teeth at posterior positions bent forward.

Xenophysogobio boulengeri **(Tchang, 1929)** (Fig. 310-9: IHCAS no number, BL 89.5 mm)

Three specimens from Sichuan, China (IHCAS no numbers). Dental formula: 3.5-5. -3; central teeth Type 4-1, anterior teeth and minor-row teeth Type 2-1. Crowns of teeth at posterior positions bent forward.

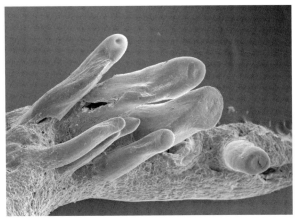

Fig. 310-1. *Gobiobotia abbreviata*: IHCAS12208010.

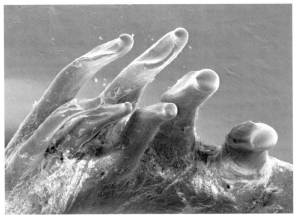

Fig. 310-2. *Gobiobotia brevibarba*: LBM1210017560.

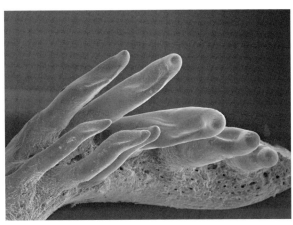

Fig. 310-3. *Gobiobotia filifer*: IHCAS no number.

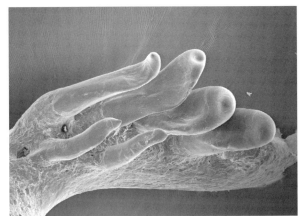

Fig. 310-4. *Gobiobotia guilingensis*: IHCAS12208060.

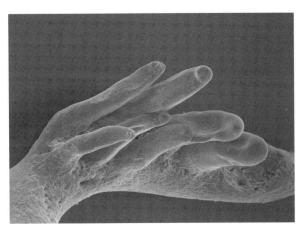

Fig. 310-5. *Gobiobotia kolleri*: IHCAS12208019.

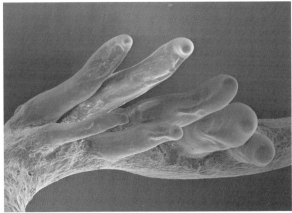

Fig. 310-6. *Gobiobotia meridionalis*: IHCAS12208018.

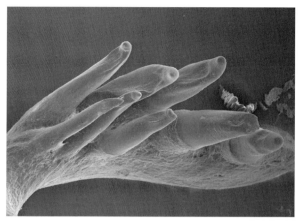

Fig. 310-7. *Gobiobotia pappenheimi*: IHCAS12208046.

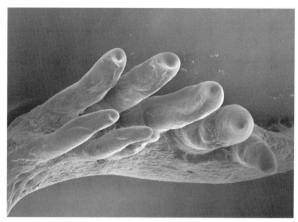

Fig. 310-8. *Gobiobotia tungi*: IHCAS12208006.

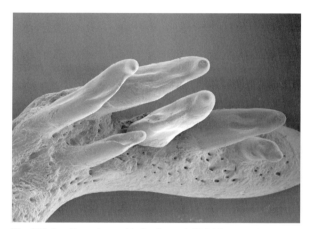

Fig. 310-9. *Xenophysogobio boulengeri*: IHCAS no number.

3-11. Gobioninae

The pharyngeal dentitions of 45 species in 20 genera of Gobioninae were examined. Their teeth are arranged in one or two rows except for *Beligobio* and *Hemibarbus*, which have three-row dentitions. The typical dental formula is 1.3.5-5.3.1 in species with three-row dentitions, 3(2,4).5(4)-5(4).3(2,4) in species with two-row dentitions, and 5-5 with no variation in species with one row of teeth. The central teeth are Type 4-1 in all species.

***Belligobio nummifer* (Boulenger, 1901)** (Fig. 311-1: IHCAS12208178, BL 75.3 mm)

Two specimens from Shaanxi, China (IHCAS12208178). Dental formula: 1.3.5-5.3.1; central teeth Type 4-1, anterior teeth and minor-row teeth Type 2-1. Crowns of posterior central and minor-row teeth bent forward. Shallow grooves present along anterior margins of central teeth.

***Belligobio pengxianensis* Luo, Le and Chen, 1977** (Fig. 311-2: IHCAS nonumber, BL 86.1 mm)

Two specimens from Sichuan, China (IHCAS no numbers). Dental formula: 1.3.5-5.(2,3).1; central teeth Type 4-1, anterior teeth and minor-row teeth Type 2-1. Crowns of posterior central and minor-row teeth bent forward. Shallow groove present along anterior margin of central teeth.

***Hemibarbus barbus* (Temminck and Schlegel, 1846)** (Fig. 311-3: LBM nonumber, BL 268.0 mm)

Eight specimens from Shiga, Japan (LBM1210003702, 1210013795-1210013798, 1210014426, 1210014427, 1210039910 and no number). Dental formula: 1.3.5-5.3.1; central teeth Type 4-1, anterior teeth Type 2-0, minor-row teeth Type 2-1. Crowns of posterior central and minor-row teeth bent forward. Shallow groove present along anterior margin of central teeth. Anterior margins of teeth A3 (An1) and A4 (Ce0) swollen. Tooth A2 (An2) massive. Some teeth worn.

***Hemibarbus labeo* (Pallas, 1776)** (Fig. 311-4: LBM1210014590, BL 285.0 mm)

Five specimens from Shiga, Japan (LBM1210014589, 1210014590, 1210024342, 1210040601, and 1210040602). Dental formula: 1.3.5-5.3.1; central teeth Type 4-1, anterior teeth Type 2-0, minor-row teeth Type 2-1. Crowns of posterior central and minor-row teeth bent forward. Shallow groove present along anterior margin of central teeth. Anterior margins of teeth A3 (An1) and A4 (Ce0) swollen. Some teeth worn.

***Hemibarbus longirostris* (Regan, 1908)** (Fig. 311-5: LBM1210014697, BL 102.8 mm)

Six specimens from Kyoto (LBM1210014694-1210014697) and Okayama, Japan (LBM1210040450 and 1210040451). Dental formula: 1.3.5-5.3.1; central teeth Type 4-1, anterior teeth Type 2-0, minor-row teeth Type 2-1. Crowns of posterior central and minor-row teeth bent forward. Shallow groove present along anterior margin of central teeth. Anterior margins of teeth A3 (An1) and A4 (Ce0) swollen. Tooth An2 (A2) massive.

***Hemibarbus maculatus* Bleeker, 1871** (Fig. 311-6: IHCAS12208118, BL 130.1 mm)

One specimen from Guangxi, China (IHCAS12208118). Dental formula: 1.3.5-5.3.1; central teeth Type 4-1, anterior teeth Type 2-0, minor-row teeth Type 2-1. Crowns of posterior central teeth and minor-row teeth bent forward. Shallow groove present along anterior margin of central teeth. Anterior margins of teeth A3 (An1) and A4 (Ce0) swollen. Tooth A2 (An2) massive.

***Acanthogobio guentheri* Herzenstein, 1892** (Fig. 311-7: IHCAS12208420, BL 108.8 mm)

Two specimens from Gansu, China (IHCAS12208420) and Jilin, China (IHCAS12208447). Dental formula: 3.5-5.3; central teeth Type 4-1, anterior teeth and minor-row teeth Type 2-1. Crowns of central teeth and posterior minor-row teeth bent forward. Shallow groove present along anterior margin of central teeth. Anterior (or medial) margins of teeth A2 (An2) and A3 (An1) swollen.

***Coreoleuciscus splendidus* Mori, 1935** (Fig. 311-8: LBM1210014984, BL 67.2 mm)

Two specimens from Korea (LBM1210014984 and 1210014985). Dental formula: 2.5-5.2; central teeth Type 4-1, anterior teeth and minor-row teeth Type 2-1. Crowns of posterior central and minor-row teeth bent forward. Several denticles present on anterior margin of central teeth.

***Gnathopogon caerulescens* (Sauvage, 1883)** (Fig. 311-9: LBM1210024319, BL 82.0 mm)

Twelve specimens from Shiga, Japan

(LBM1210013776, 1210024319, 1210024331, 1210024333, 1210040494, 1210040496, 1210040498, 1210040500, 1210040502, and 1210040504- 1210040506). Dental formula: 2(3).5-5.2(3); central teeth Type 4-1, anterior teeth and minor-row teeth Type 2-1. Crowns of posterior central and minor-row teeth bent forward. Many denticles present on grinding surface of central teeth.

Gnathopogon elongatus (**Temminck & Schlegel, 1846**) (Fig. 311-10: LBM1210013735, BL 78.8 mm)

Nine specimens from Gifu (LBM1210013742, 1210013744, 1210013746,), Shiga, (LBM1210013721- 1210013723 and 1210013748), and Saga, Japan (LBM1210013734 and 1210013735). Dental formula: 2(3).5-5(4).2(3); central teeth Type 4-1, anterior teeth and minor-row teeth Type 2-1 or 2-0. Crowns of posterior central and minor-row teeth bent forward. Several denticles present on anterior margin of central teeth.

Gnathopogon herzensteini (**Günther, 1896**) (Fig. 311-11: IHCAS12208483, BL 97.0 mm)

Two specimens from Henan, China (IHCAS12208483). Dental formula: 3.5-5.3; central teeth Type 4-1, anterior teeth and minor-row teeth Type 2-1. Crowns of posterior central and minor-row teeth bent forward. Anterior (or lateral) margins of teeth A2 (An2) and A3 (An1) swollen. Tooth A2 (An2) massive. Some teeth worn.

Gnathopogon imberbis (**Sauvage & Dabry de Thiersant, 1874**) (Fig. 311-12: LBM1210014631, BL 95.5 mm)

Three specimens from Gansu, China (IHCAS12208491, LBM1210014630, and 1210014631). Dental formula: 2(4).5(4)-5(4).(2,3,4); central teeth Type 4-1, anterior teeth and minor-row teeth Type 2-1. Crowns of posterior central and minor-row teeth bent forward. Shallow groove present along anterior margin of central teeth.

Gnathopogon nicholsi (**Fang, 1943**) (Fig. 311-13: IHCAS12208466, BL 87.0 mm)

One specimen from Jiangxi, China (IHCAS12208466). Dental formula: 3.5-5.3; central teeth Type 4-1, anterior teeth and minor-row teeth Type 2-1 or 2-0. Crowns of posterior central and minor-row teeth bent forward. Anterior margin of tooth A3 (An1) swollen. Some teeth worn.

Gnathopogon taeniellus (**Nichols, 1925**) (Fig. 311-14: IHCAS12208473, BL 70.1 mm)

Two specimens from Zhejiang, China (IHCAS12208473). Dental formula: (2,3).5-5.(2,3); central teeth Type 4-1, anterior teeth and minor-row teeth Type 2-1. Crowns of posterior central and minor-row teeth bent forward. Shallow groove present along anterior margin of central teeth. Anterior margin of tooth A3 (An1) swollen. Some teeth worn.

Gnathopogon tsinanensis (**Mori, 1928**) (Fig. 311-15: IHCAS12208468, BL 79.9 mm)

Three specimens from Shannxi, China (IHCAS12208468). Dental formula: 3.5(4)-4(5).3(2); central teeth Type 4-1, anterior teeth Type 2-0, minor-row teeth Type 2-1 or 2-0. Crowns of posterior central and minor-row teeth bent forward. Shallow groove present along anterior margin of central teeth. Anterior margin of teeth A3 (An1) and A4 (Ce0) swollen.

Paracanthobrama guichenoti **Bleeker, 1864** (Fig. 311-16: IHCAS12208420, BL 106.2 mm)

Two specimens from Jiangxi, China (IHCAS12208420). Dental formula: 4.5-5.4; central teeth Type 4-1, anterior teeth Type 4-1 or 2-0, minor-row teeth Type 2-1. Crowns of posterior central and minor-row teeth bent forward. Anterior margins of teeth A3 (An1) and A4 (Ce0) swollen. Some teeth worn.

Rhinogobio ventralis **Sauvage and Dabry de Thiersant, 1874** (Fig. 311-17: IHCAS12208693, BL 182.2 mm)

Two specimens from Sichuan, China (IHCAS12208693). Dental formula: 2.5-5.2; central teeth Type 4-1, anterior teeth Type 2-0, minor-row teeth Type 2-1 or 2-0. Crowns of posterior central and minor-row teeth bent forward. Anterior margins of teeth A3 (An1) and A4 (Ce0) swollen. Some teeth worn.

Squalidus chankaensis chankanensis **Dybowsky, 1872** (Fig. 311-18: LBM1210024460, BL 32.5 mm)

Six specimens from Okayama (LBM1210024460) and Aichi, Japan (LBM1210040569, 1210040571, 1210040573, 1210040575, and 1210047576). Dental formula: 3(2).5-5.3(2); central teeth Type 4-1, anterior teeth and minor-row teeth Type 2-1. Crowns of central teeth and minor-row teeth bent forward. Shallow groove present along anterior margin of central teeth. Anterior margins of teeth A3 (An1) and A4 (Ce0) swollen.

Squalidus chankaensis biwae (Jordan and Snyder, 1900) (Fig. 311‒19: LBM1210014645, BL 66.6 mm)

Four specimens from Shiga, Japan (LBM1210014644‒1210014647). Dental formula: 3(2).5‒5.3; central teeth Type 4‒1, anterior teeth and minor-row teeth Type 2‒1. Crowns of posterior central and minor-row teeth bent forward. Shallow groove present along anterior margin of central teeth. Anterior margins of teeth A3 (An1) and A4 (Ce0) swollen.

Squalidus gracilis gracilis (Temminck and Schlegel, 1846) (Fig. 311‒20: LBM1210014420, BL 53.7 mm)

Five specimens from Shiga, Japan (LBM1210003615 and 1210014420‒1210014423). Dental formula: 3(2).5‒5.3(2); central teeth Type 4‒1, anterior teeth and minor-row teeth Type 2‒1. Crowns of posterior central and minor-row teeth bent forward. Anterior margin of tooth A3 (An1) swollen. Some tooth well worn.

Squalidus japonicus japonicus (Sauvage, 1883) (Fig. 311‒21: LBM1210003616, BL 77.6 mm)

Ten specimens from Shiga, Japan (LBM1210003616‒1210003618, 1210040588, 1210040590, 1210040592, 1210040594, 1210040596, 1210040598, and 1210040600). Dental formula: 3.5‒5.3(2); central teeth Type 4‒1, anterior teeth and minor-row teeth Type 2‒1. Crowns of posterior central and minor-row teeth bent forward. Anterior margin of tooth A3 (An1) swollen.

Ladislavia taczanowski Dybowsky, 1869 (Fig. 311‒22: IHCAS12208295, BL 47.2 mm)

Three specimens from Heilonjiang, China (IHCAS12208295). Dental formula: 2.5‒5(4).2(3); central teeth Type 4‒1, anterior teeth Type 4‒1 or 2‒1, minor-row teeth Type 2‒1. Crowns of central and minor-row teeth bent forward. Shallow groove present along anterior margin of central teeth. Some teeth well worn, with secondary grinding surface.

Pseudogobio esocinus (Temminck and Schlegel, 1846) (Fig. 311‒23: LBM1210029388, BL 82.1 mm)

Seven specimens from Shiga, Japan (LBM1210029328, 1210029344, 1210029348, 1210029352, 1210029354, 1210029378, and 1210029388). Dental formula: 2.5‒5.2(3); major-row teeth Type 4‒1, minor-row teeth Type 2‒1. Crowns of posterior central and minor-row teeth bent forward. Dentigerous surface somewhat produced medially, with central teeth

arising from anterior side of spur. Crowns of posterior central teeth compressed antero-posteriorly.

Pseudogobio vaillanti (Sauvage, 1878) (Fig. 311‒24: IHCAS12208860, BL 87.0 mm)

Two specimens from Liaoning, China (IHCAS12208860). Dental formula: 5‒5; central teeth Type 4‒1, anterior teeth Type 2‒3 or 2‒1, minor-row teeth Type 2‒1. Crowns of posterior minor-row teeth bent forward. Dentigerous surface somewhat produced medially, with central teeth arising from anterior side of spur. Crowns of central teeth compressed antero-posteriorly.

Sarcocheilichthys davidi (Sauvage, 1878) (Fig. 311‒25: IHCAS12208355, BL 60.0 mm)

Two specimens from Sichuan, China (IHCAS12208355). Dental formula: 1.5‒5.1; central teeth Type 4‒1, anterior teeth and minor-row teeth Type 2‒1. Dentigerous surface somewhat produced medially, with central teeth arising from anterior side of spur. Crowns of teeth A4 (Ce0) and A5 (Po1) compressed antero-posteriorly. Tooth A3 (An1) with swollen anterior margin and shallow groove on grinding surface.

Sarcocheilichthys parvus Nikols, 1930 (Fig. 301‒26: LBM1210014466, BL 44.0 mm)

Four specimens from Jiangsu, China (LBM1210014464‒1210014467). Dental formula: 5‒5; all teeth Type 4‒1. Dentigerous surface somewhat produced medially, with central teeth arising from anterior side of spur. Crowns of posterior central teeth compressed antero-posteriorly.

Sarcocheilichthys sinensis Bleeker, 1871 (Fig. 301‒27: LBM1210014434, BL 75.0 mm)

Six specimens from Jiangsu, China (LBM1210014433, 1210014434, 1210014437, 1210014438, 1210014441, and 1210014442). Dental formula: 5‒5; central teeth Type 4‒1, anterior teeth Type 4‒1 or 2‒0. Crowns of posterior central teeth and minor-row teeth bent forward. Dentigerous surface somewhat produced medially, with central teeth arising from anterior side of spur. Shallow groove present along anterior margin of central teeth. Anterior margins of teeth A2 (An2) to A4 (Ce0) swollen.

Sarcocheilichthys biwaensis **Hosoya, 1982** (Fig. 311‒28: LBM1210003626, BL 104.0 mm)

Four specimens from Shiga, Japan (LBM1210003626, 1210013945, 1210014346, and 1210014347). Dental formula: 5‒5; central teeth Type 4‒1, anterior teeth Type 2‒1. Dentigerous surface somewhat produced medially, with central teeth arising from anterior side of spur. Shallow groove present along anterior margin of central teeth. Tooth A5 (Po1) compressed antero-posteriorly. Anterior margins of teeth A3 (An1) and A4 (Ce0) swollen. Tooth A2 (An2) massive.

Sarcocheilichthys variegatus variegatus (**Temminck and Schlegel, 1846**) (Fig. 311‒29: LBM 1210014335, BL 82.9 mm)

Three specimens from Hyogo, Japan (LBM1210014333‒1210014335). Dental formula: 5‒5; central teeth Type 4‒1, anterior teeth Type 2‒1. Dentigerous surface somewhat produced medially, with central teeth arising from anterior side of spur. Shallow groove present along anterior margin of central teeth. Tooth A5 (Po1) compressed antero-posteriorly. Anterior margins of teeth A3 (An1) and A4 (Ce0) swollen. Tooth A2 (An2) massive.

Sarcocheilichthys variegatus microoculis **Mori, 1927** (Fig. 311‒30: LBM1210003621, BL 109.2 mm)

Seven specimens from Shiga, Japan (LBM1210003619‒1210003625). Dental formula: 5‒5; central teeth Type 4‒1, anterior teeth Type 2‒1. Dentigerous surface somewhat produced medially, with central teeth arising from anterior side of spur. Shallow groove present along anterior margin of central teeth. Anterior margins of teeth A3 (An1) and A4 (Ce0) swollen. Tooth A2 (An2) massive.

Saurogobio immaculatus **Koller, 1927** (Fig. 311‒31: IHCAS122081077, BL 132.0 mm)

Two specimens from Hainan, China (IHCAS122081077). Dental formula: 5‒5; central teeth Type 4‒1, anterior teeth Type 4‒1 or 2‒1. Dentigerous surface somewhat produced medially, with central teeth arising from anterior side of spur. Posterior central teeth compressed antero-posteriorly. Tooth crowns well worn with secondary grinding surfaces.

Abbotina rivularis (**Basilewsky, 1855**) (Fig. 311‒32: LBM1210003692, BL 73.8 mm)

Eight specimens from Jiangxi, China (IHCAS12208817), Fukuoka (LBM1210003691‒1210003693, and 1210013821), and Shiga, Japan (LBM1210043713, 1210043833, and 1210043834). Dental formula: 5‒5; central teeth Type 4‒1, anterior teeth Type 4‒1 or 2‒3. Dentigerous surface strongly produced medially, with central teeth arising from anterior side of spur. Tooth crowns well worn with secondary grinding surfaces.

Pseudorasbora elongata **Wu, 1939** (Fig. 311‒33: IHCAS12208292, BL 115.0 mm)

One specimen from Anhui, China (IHCAS12208292). Dental formula is 5‒5; central teeth Type 4‒1, anterior teeth Type 4‒1 or 2‒0. Dentigerous surface strongly produced medially, with central teeth arising from anterior side of spur. Some teeth worn.

Pseudorasbora parva (**Temminck and Schlegel, 1846**) (Fig. 311‒34: LBM1210003645, BL65.0 mm)

Nine specimens from Shiga, Japan (LBM1210003643‒1210003647 and 1210014404‒1210014407). Dental formula: 5‒5; central teeth Type 4‒1 or 4‒0, anterior teeth Type 4‒1 or 2‒3. Dentigerous surface strongly produced medially, with central teeth arising from anterior side of acute spur. Crowns of tooth A5 (Po1) compressed antero-posteriorly. Denticles present on both margins of uneroded teeth. Some teeth worn.

Pseudorasbora pumila pumila **Miyadi, 1930** (Fig. 311‒35: LBM1210014352, BL 54.0 mm)

Four specimens from Niigata, Japan (LBM1210014352‒1210014355). Dental formula: 5‒5; all teeth Type 4‒1. Dentigerous surface strongly produced medially, with central teeth arising from anterior side of acute spur. Crown of tooth A5 (Po1) compressed antero-posteriorly. Denticles present on both margins of uneroded teeth. Some teeth worn.

Pseudorasbora pumila **subsp. A,** Japanese name: ushimotsugo (Fig. 311‒36: LBM1210014359, BL 46.0 mm)

Five specimens from Aichi, Japan (LBM1210014358, 1210014359, 1210040044, 1210040046, 1210040048). Dental formula: 5‒5; all teeth Type 4‒1. Dentigerous surface strongly produced medially, with central teeth arising from anterior side of acute spur.

Coreius guichenoti (**Sauvage and Dabry de Thiersant, 1874**) (Fig. 311-37: IHCAS12208613, BL 110.9 mm)

Two specimens from Hubei, China (IHCAS12208613). Dental formula: 5-5; central teeth Type 4-1, anterior teeth Type 2-1 or 2-0. Crowns of posterior central teeth at posterior positions bent forward. Wide, shallow groove present along anterior margin of teeth A4 (Ce0) and A5 (Po1).

Coreius heterodon (**Bleeker, 1864**) (Fig. 311-38: IHCAS12208646, BL 91.9 mm)

Three specimens from Hubei, China (IHCAS12208646). Dental formula: 5-5; central teeth Type 4-1, anterior teeth Type 4-1 or 2-0. Crowns of posterior central teeth bent forward. Wide, shallow groove present along anterior margin of tooth A5 (Po1).

Biwia zezera (**Ishikawa, 1895**) (Fig. 311-39: LBM1210013910, BL 50.6 mm)

Twelve specimens from Shiga, Japan (LBM1210003715, 1210003716, 1210013816, 1210013817, 1210013909, 1210013909-1210013911, 1210041489, 1210041491, 1210041493, 1210041495, and 1210041497). Dental formula: 5-5; all teeth Type 4-1. Dentigerous surface strongly produced medially, with central teeth arising from anterior side of acute spur. Hamulus groove present on posterior side of tooth crowns. Tooth crowns compressed antero-posteriorly.

Microphysogobio brevirostris (**Günther, 1868**) (Fig. 311-40: LBM1210015025, BL 67.5 mm)

One specimen from Taiwan, China (LBM1210015025). Dental formula: 5-5; all teeth Type 4-1. Dentigerous surface strongly produced medially, with central teeth arising from anterior side of acute spur. Tooth crowns compressed antero-posteriorly.

Microphysogobio tungtingensis (**Nikols, 1926**) (Fig. 311-41: LBM 1210013837, BL 76.0 mm)

One specimen from Korea (LBM 1210013837). Dental formula: 5-5; all teeth Type 4-1. Dentigerous surface strongly produced medially, with central teeth arising from anterior side of acute spur. Tooth crowns compressed antero-posteriorly.

Microphysogobio labeoides (**Nikols and Pope, 1927**) (Fig. 311-42, IHCAS12208927, BL 73.9 mm)

Two specimens from Hainan, China (IHCAS12208927). Dental formula: 5-5; all teeth Type 4-1. Dentigerous surface strongly produced medially, with central teeth arising from anterior side of spur. Tooth crowns compressed antero-posteriorly, well worn with secondary grinding surfaces.

Microphysogobio amurensis (**Taranetz, 1937**) (Fig. 311-43: IHCAS12208383, BL 46.7 mm)

Three specimens from Heilongjiang, China (IHCAS12208383). Dental formula: 5-5; all teeth Type 4-1. Dentigerous surface strongly produced medially, with central teeth arising from anterior side of acute spur. Tooth crowns compressed antero-posteriorly, well worn with secondary grinding surfaces.

Platysmacheilus exiguous (**Lin, 1932**) (Fig. 311-44: IHCAS12208706, BL 71.6 mm)

Two specimens from Shaanxi, China (IHCAS12208706). Dental formula: 5-5; all teeth Type 4-1. Dentigerous surface strongly produced medially, with central teeth arising from anterior side of spur. Tooth crowns compressed antero-posteriorly.

Pungtungia herzi **Herzenstein, 1892** (Fig. 311-45: LBM1210003673, BL97.0 mm)

Six specimens from Kyoto, Japan (LBM1210003672-1210003677). Dental formula: 5-5; central teeth Type 4-1, anterior teeth Type 4-1 or 2-0. Dentigerous surface strongly produced medially, with central teeth arising from anterior side of acute spur. Shallow groove present along anterior margin of central teeth. Tooth A5 (Po1) compressed. Tooth A1 (An3) massive.

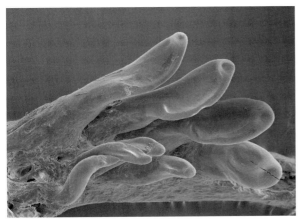

Fig. 311-1. *Belligobio nummifer*: IHCAS12208178.

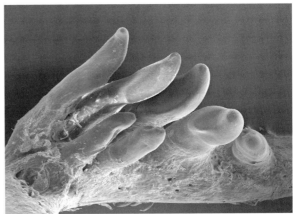

Fig. 311-2. *Belligobio pengxianensis*: IHCAS no number.

Fig. 311-3. *Hemibarbus barbus*: LBM no number.

Fig. 311-4. *Hemibarbus labeo*: LBM1210014590.

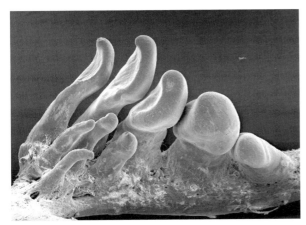

Fig. 311-5. *Hemibarbus longirostris*: LBM1210014697.

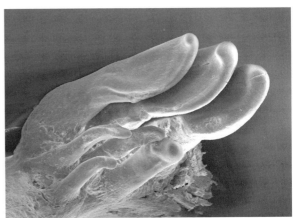

Fig. 311-6. *Hemibarbus maculatus*: IHCAS12208118.

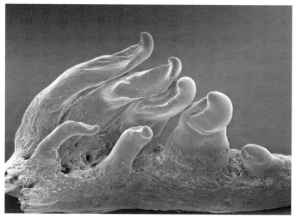

Fig. 311-7. *Acanthogobio guentheri*: IHCAS12208420.

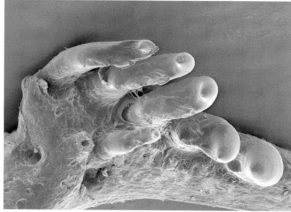

Fig. 311-8. *Coreoleuciscus splendidus*: LBM1210014984.

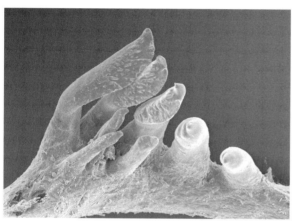

Fig. 311-9. *Gnathopogon caerulescens*: LBM1210024319.

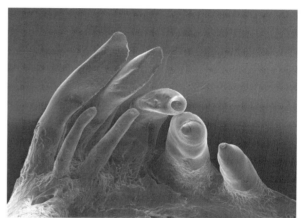

Fig. 311-10. *Gnathopogon elongatus*: LBM1210013735.

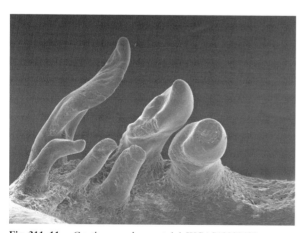

Fig. 311-11. *Gnathopogon herzensteini*: IHCAS12208483.

Fig. 311-12. *Gnathopogon imberbis*: LBM1210014631.

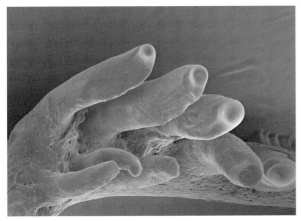

Fig. 311-13. *Gnathopogon nicholsi*: IHCAS12208466.

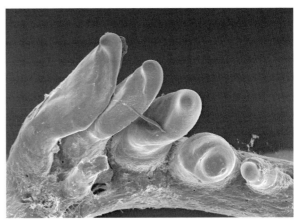

Fig. 311-14. *Gnathopogon taeniellus*: IHCAS12208473.

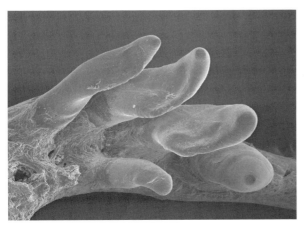

Fig. 311-15. *Gnathopogon tsinanensis*: IHCAS12208468.

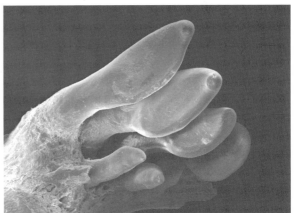

Fig. 311-16. *Paracanthobrama guichenoti*: IHCAS12208420.

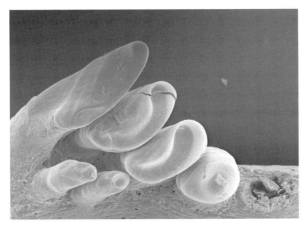

Fig. 311-17. *Rhinogobio ventralis*: IHCAS12208693.

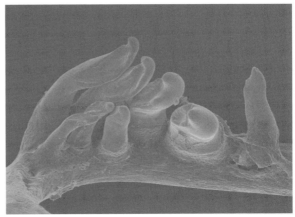

Fig. 311-18. *Squalidus chankaensis chankanensis*: LBM1210024460.

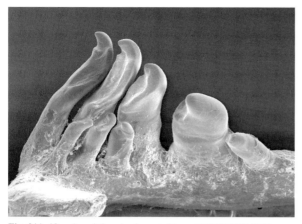

Fig. 311-19. *Squalidus chankaensis biwae*: LBM1210014645.

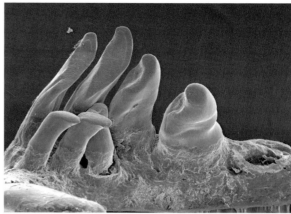

Fig. 311-:20. *Squalidus gracilis gracilis*: LBM1210014420.

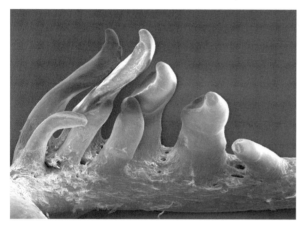

Fig.311-21. *Squalidus japonicus japonicus*: LBM1210003616.

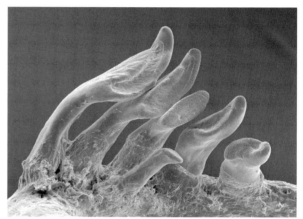

Fig. 311-22. *Ladislavia taczanowski*: IHCAS12208295.

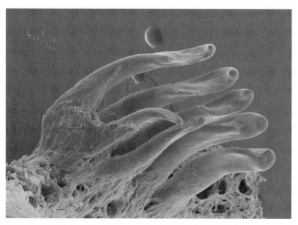

Fig. 311-23. *Pseudogobio esocinus*: LBM1210029388.

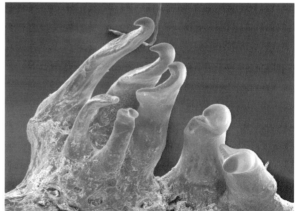

Fig. 311-24. *Pseudogobio vaillanti*: IHCAS12208860.

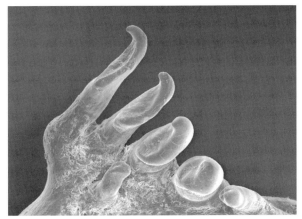

Fig. 311-25. *Sarcocheilichthys davidi*: IHCAS12208355.

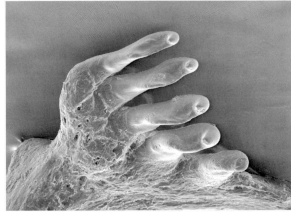

Fig. 301-26. *Sarcocheilichthys parvus*: LBM1210014466.

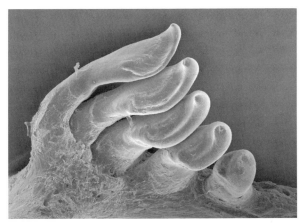

Fig. 301-27. *Sarcocheilichthys sinensis*: LBM1210014434.

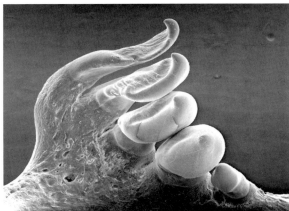

Fig. 311-28. *Sarcocheilichthys biwaensis*: LBM1210003626.

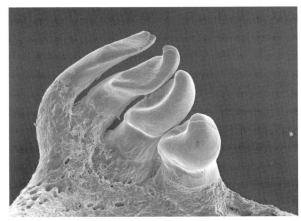

Fig. 311-29. *Sarcocheilichthys variegatus variegatus*: LBM 1210014335.

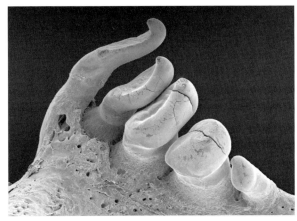

Fig. 311-30. *Sarcocheilichthys variegatus microoculis*: LBM1210003621.

3. Descriptions of the pharyngeal dentition of cyprinid subfamilies

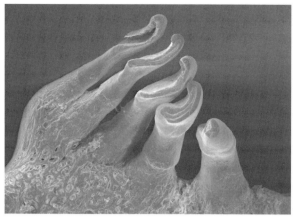

Fig. 311-31. *Saurogobio immaculatus*: IHCAS122081077.

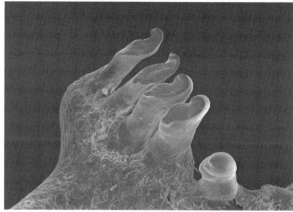

Fig. 311-32. *Abbotina rivularis*: LBM1210003692.

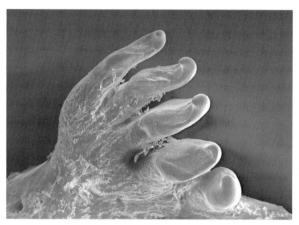

Fig. 311-33. *Pseudorasbora elongata*: IHCAS12208292.

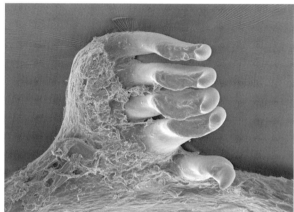

Fig. 311-34. *Pseudorasbora parva*: LBM1210003645.

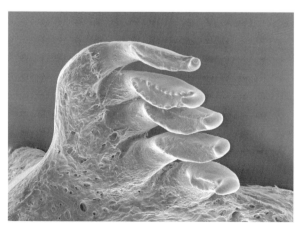

Fig. 311-35. *Pseudorasbora pumila pumila*: LBM1210014352.

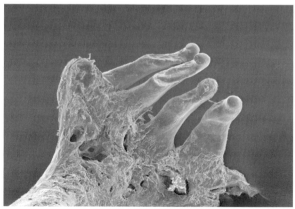

Fig. 311-36. *Pseudorasbora pumila* subsp. A: LBM1210014359.

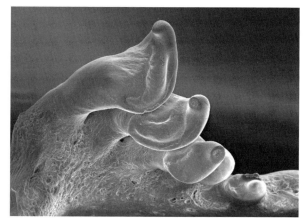

Fig. 311-37. *Coreius guichenoti*: IHCAS12208613.

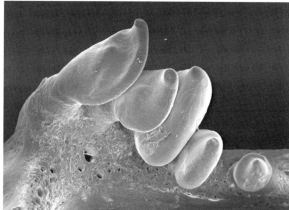

Fig. 311-38. *Coreius heterodon*: IHCAS12208646.

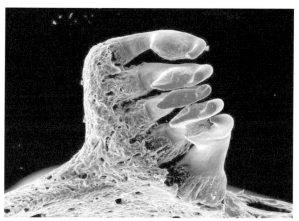

Fig. 311-39. *Biwia zezera*: LBM1210013910.

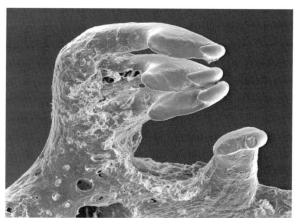

Fig. 311-40. *Microphysogobio brevirostris*: LBM1210015025.

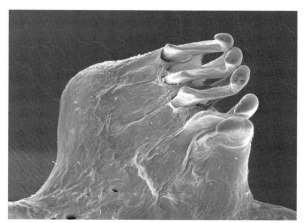

Fig. 311-41. *Microphysogobio tungtingensis*: LBM 1210013837.

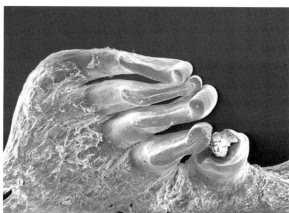

Fig. 311-42. *Microphysogobio labeoides*:IHCAS12208927.

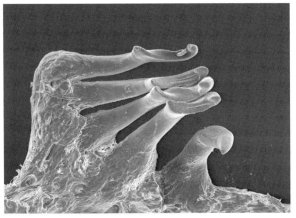

Fig. 311-43. *Microphysogobio amurensis*: IHCAS12208383.

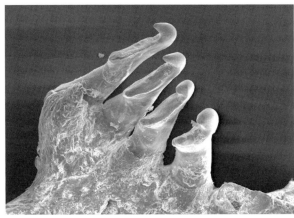

Fig. 311-44. *Platysmacheilus exiguous*: IHCAS12208706.

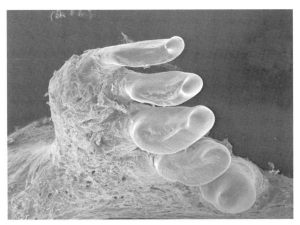

Fig. 311-45. *Pungtungia herzi*: LBM1210003673.

3-12. Cyprininae.

The pharyngeal dentitions of 22 species in five genera of Cyprininae were examined. The central teeth in *Puntioplithes* and *Procypris* are Type 2-2, and their dental formula is 2.3.4(5)-4(5).3.2, just as in Barbinae. Although the central teeth are generally Type 5 in *Cyprinus*, *Carassioides*, and *Carassius*, the most posterior central tooth (tooth A3: position Ce0) is Type 4-1 in some species of *Cyprinus*. The teeth are arranged in three rows in *Cyprinus*, two rows in *Carassioides*, and one row in *Carassius*, and their dental formulae are 1.1.3-3.1.1, 2.4-4.2, and 4-4, respectively. Variation in tooth number hardly occurs.

Puntioplites bulu (Bleeker, 1851) (Fig. 312-1: LBM 1210024413, BL 52.0 mm)

One specimen from Malaysia (LBM 1210024413). Dental formula: 2.3.5-5.3.2; central teeth and minor-row teeth Type 2-2, anterior teeth Type 2-1 or 2-0. Crowns of central teeth and minor-row teeth bent forward with grinding surface oriented dorsad. Longitudinal ridge on grinding surface of posterior central teeth and minor-row teeth.

Puntioplites facifer Smith, 1929 (Fig. 312-2: LBM1210047495, BL 58.7 mm)

One specimen from Cambodia (LBM1210047495). Dental formula: 2.3.4-5.3.2; central teeth and minor-row teeth Type 2-2, anterior teeth Type 2-2 or 2-1. Crowns of central teeth and minor-row teeth bent forward with grinding surface oriented dorsad. Longitudinal ridge on grinding surface of posterior central teeth and minor-row teeth. Tips of both margins of grinding surfaces swollen into humps.

Puntioplites proctozystron (Bleeker, 1865) (Fig. 312-3: LBM1210013988, BL 74.0 mm)

Two specimens from Thailand (LBM1210013988 and 1210014065). Dental formula: 2.3.4-4.3.2; central teeth and minor-row teeth Type 2-2, anterior teeth Type 2-2 or 2-1. Crowns of central teeth and minor-row teeth bent forward with grinding surface oriented dorsad. Longitudinal ridge on grinding surface of posterior central teeth and minor-row teeth. Tips of both margins of grinding surfaces swollen into humps.

Puntioplites waandersi (Bleeker, 1859) (Fig. 312-4: LBM1210014060, BL 136.5 mm)

One specimens from Thailand (LBM1210014060). Dental formula: 2.3.4-4.3.2; central teeth and minor-row teeth Type 2-2, anterior teeth Type 2-2 or 2-1. Crowns of central teeth and minor-row teeth bent forward with grinding surface oriented dorsad. Longitudinal ridge on grinding surface of posterior central and minor-row teeth. Tips of both margins of grinding surfaces swollen into humps.

Procypris rabaudi (Tchang, 1930) (Fig. 312-5: IHCAS no number, BL 167.0 mm)

One specimen from Sichuan, China (IHCAS no number). Dental formula: 2.3.4-4.3.2; central teeth Type 2-2, anterior teeth Type 2-1, minor-row teeth Type 2-2 or 2-1. Crowns of posterior central and minor-row teeth bent forward with grinding surface oriented dorsad. U-shaped groove on grinding surface of posterior central and minor-row teeth. Some teeth well worn with secondary grinding surface.

Cyprinus multitaeniatus Pellegrin and Chevey, 1936 (Fig. 312-6: IHCAS589988, BL 25.5 mm)

One specimens from Guangxi, China (IHCAS589988). Dental formula: 1.1.3-3.1.1; central teeth Type 4-1 or 5-1, anterior teeth Type 2-0, minor-row teeth Type 5-1. Single transverse groove present on grinding surface of central teeth and minor-row teeth. Tooth A1 (An2) massive.

Cyprinus yilongensis Yang *et al.*, 1977 (Fig. 312-7: IHCAS12212032, BL 90.4 mm)

One specimen from Yunnan, China (IHCAS12212032). Dental formula: 1.1.3-3.1.1; central teeth Type 4-1 or 5-1, anterior teeth Type 2-0, minor-row teeth Type 5-1. Single transverse groove present on grinding surface of central teeth and minor-row teeth. Tooth A1 (An2) massive.

Cyprinus barbatus Chen and Huang, 1977 (Fig. 312-8: IHCAS12212144, BL 394. 0 mm)

One specimen from Yunnan, China (IHCAS12212144). Dental formula: 1.1.3-3.1.1; central teeth Type 4-1 or 5-1, anterior teeth Type 2-0, minor-row teeth Type 5-1. Multiple transverse grooves present on grinding surface of central teeth and minor-row teeth. Tooth A1 (An2) massive.

Cyprinus carpio **Linnaeus, 1758** (Fig. 312-9: LBM1210032994, BL 420.0 mm)

104 specimens from Shiga, Japan (LBM1210013611, 1210013661, 1210013737, 1210013739, 1210013741, 1210013788-1370013791, 1210013944, 1210021700, 1210031650, 1210031652, 1210031654, 1210031656, 1210031658, 1210031660, 1210031662, 1210031664, 1210031666, 1210031770, 1210032045, 12100332047, 1210032986-1210032990, 1210033972-1210033980, 1210036819, 1210039922, 1210046425, 1210046427, 1210046429, 1210046431, 1210046433, 1210046435, 1210046437, 1210046439, 1210046441, 1210046443, 1210046445, 1210046462-1210046464, 1210046503-1210046552). Dental formula: 1.1.3-3.1.1; central teeth and minor-row teeth Type 5-1, anterior teeth Type 2-0. Multiple transverse grooves present on grinding surface of central teeth and minor-row teeth. Tooth A1 (An2) massive.

Cyprinus chilia **Wu, Yang and Huang, 1963** (Fig. 312-10: IHCAS12212120, BL 68.7 mm)

Two specimens from Yunnan, China (IHCAS12212120). Dental formula: 1.1.3-3.1.1; central teeth and minor-row teeth Type 5-1, anterior teeth Type 2-1. Multiple transverse grooves present on grinding surface of tooth A2 (An1). Tooth A1 (An2) massive and bearing pointed tooth hook. Some teeth well worn.

Cyprinus longipectoralis **Chen and Huang, 1977** (Fig. 312-11: IHCAS12212139, BL 182.2 mm)

Two specimens from Yunnan, China (IHCAS12212139 and 12212120). Dental formula: 1.1.3-3.1.1; central teeth and minor-row teeth Type 5-1, anterior teeth Type 2-0. Multiple transverse grooves present on grinding surface of tooth A2 (An1). Tooth A1 (An2) massive and bearing pointed tooth hook. Some teeth well worn.

Cyprinus pellegrini **Tchang, 1933** (Fig. 312-12: IHCAS12212141, BL 269.0 mm)

Three specimens from Yunnan, China (IHCAS12212141). Dental formula: 1.1.3-3.1.1; central teeth and minor-row teeth Type 5-1, anterior teeth Type 2-0. Multiple transverse grooves present on grinding surface of central teeth and minor-row teeth. Tooth A1 (An2) massive.

Cyprinus rubrofuscus **Lacepède, 1803** (Fig. 312-13: IHCAS122112103, BL 224.9 mm)

Two specimens from Hainan, China (IHCAS122112103). Dental formula: 1.1.3-3.1.1; central teeth and minor-row teeth Type 5-1, anterior teeth Type 2-0. Multiple transverse grooves present on grinding surface of central teeth and minor-row teeth. Tooth A1 (An2) massive.

Cyprinus yunnanensis **Tchang, 1933** (Fig. 312-14: IHCAS1210016524, BL 167.9 mm)

Three specimens from Yunnan, China (IHCAS12212149). Dental formula: 1.1.3-3.1.1; central teeth Type 4-1 or 5-1, anterior teeth Type 2-0, minor-row teeth Type 5-1. Multiple transverse grooves present on grinding surface of tooth A2 (An1). Tooth A1 (An2) massive and bearing pointed tooth hook. Some teeth worn.

Carassioides acuminatus **(Richardson, 1846)** (Fig. 312-15: IHCAS12212160, BL 144.2 mm)

Two specimens from Guangdong, China (IHCAS12212160). Dental formula: 2.4-4.2; central teeth Type 5-2, anterior teeth and minor-row teeth Type 2-0. Central teeth compressed antero-posteriorly. Minor-row teeth slender and rod-like. Teeth well worn with secondary grinding surfaces.

Carassius carassius **(Linnaeus, 1758)** (Fig. 312-16: LBM1210013782, BL 69.3 mm)

Four specimens from Netherlands (LBM1210013782-1210013785). Dental formula: 4-4; central teeth Type 5-2 or 5-0, anterior teeth Type 2-0. Posterior central teeth compressed antero-posteriorly. Tooth A2 (An2) and A3 (An1) with transverse shallow groove.

Carassius auratus **subsp. A,** Japanese name: Kinbuna (Fig. 312-17: LBM 1210042630, BL 66.5 mm)

Three specimens from Miyagi (LBM1210042630) and Fukushima, Japan (LBM1210042632 and 1210042634). Dental formula: 4-4; central teeth Type 5-2, anterior teeth Type 2-0. Central teeth compressed antero-posteriorly. Tooth crowns well worn with secondary grinding surfaces.

Carassius auratus buergeri **Temminck and Schlegel, 1846** (Fig. 312-18: NSM-P61704-5, BL 101.3 mm)

Five specimens from Kochi, Japan (NMS-P61704).

Dental formula: 4–4; central teeth Type 5–2, anterior teeth Type 2–0. Central teeth compressed antero-posteriorly. Tooth crowns well worn with secondary grinding surfaces.

***Carassius auratus* subsp. B,** Japanese name: Nagabuna (Fig. 312–19: NSM-PSK1350–1, BL 106.5 mm)

Three specimens from Nagano, Japan (NSM-PSK1350). Dental formula: 4–4; central teeth Type 5–2, anterior teeth Type 2–0. Central teeth compressed antero-posteriorly. Tooth crowns well worn with secondary grinding surfaces.

***Carassius auratus grandoculis* Temminck and Schlegel, 1846** (Fig. 312–20: LBM1210013627, BL 203.8 mm)

89 specimens from Shiga, Japan (LBM1210003795–1210003826, 1210013626–1210013628, 1210013632, 1210013633, 1210032976, 1210040444, 1210046723, 1210047642–1210047668, 1210049534, 1210049536, 1210049546–1210049554, 1210049563–1210049571). Dental formula: 4–4; central teeth Type 5–2, anterior teeth Type 2–0. Central teeth compressed antero-posteriorly. Tooth crowns well worn with secondary grinding surfaces.

***Carassius auratus langsdorfii* Temminck and Schlegel, 1846** (Fig. 312–21: LBM1210013630, BL 180.5 mm)

50 specimens from Shiga, Japan (LBM1210003763–1210003794, 1210013629–1210013631, 1210013682–1210013692, 1210013738, 1210032998, 1210039918, 1210039826, 1210039828, 1210039919, 1210048101, 1210043621, 1210043623,1210043630–1210043632, 1210047524–1210047562, 1210049555–1210049562, 1210049790–1210049802). Dental formula: 4–4; central teeth Type 5–2, anterior teeth Type 2–0. Central teeth compressed antero-posteriorly. Tooth crowns well worn with secondary grinding surfaces.

***Carassius cuvieri* Temminck and Schlegel, 1846** (Fig. 312–22: LBM1210013620, BL 183.5 mm)

77 specimens from Shiga, Japan (LBM1210003721–1210003762, 1210013620–1210013625, 1210013091, 1210032999, 1210047563–1210047576, 121001210049535, 1210049537–121004954, 1210049803, 1210049804). Dental formula: 4–4; central teeth Type 5–2, anterior teeth Type 2–0. Central teeth compressed antero-posteriorly. Tooth crowns well worn with secondary grinding surfaces. Enameloid layer very thin.

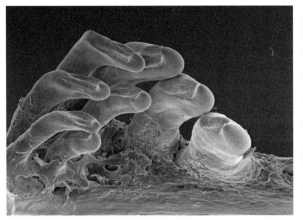

Fig. 312-1. *Puntioplites bulu*: LBM 1210024413.

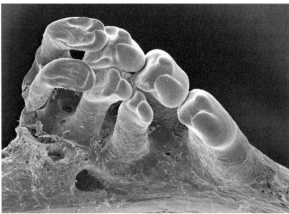

Fig. 312-2. *Puntioplites facifer*: LBM1210047495.

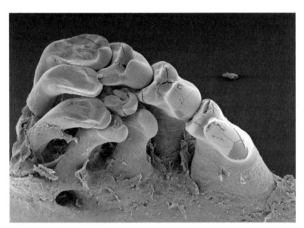

Fig. 312-3. *Puntioplites proctozystron*: LBM1210013988.

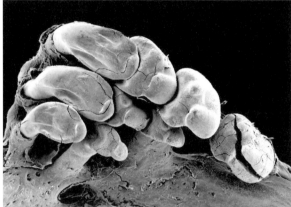

Fig. 312-4. *Puntioplites waandersi*: LBM1210014060.

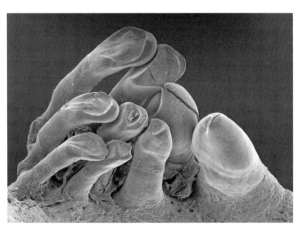

Fig. 312-5. *Procypris rabaudi*: IHCAS no number.

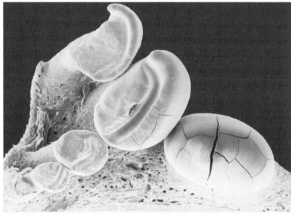

Fig. 312-6. *Cyprinus multitaeniata*: IHCAS589988.

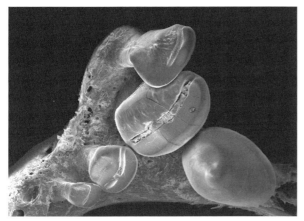

Fig. 312-7. *Cyprinus yilongensis*: IHCAS12212032.

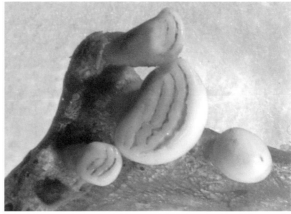

Fig. 312-8. *Cyprinus barbatus*: IHCAS12212144.

Fig. 312-9. *Cyprinus carpio*: LBM1210032994.

Fig. 312-10. *Cyprinus chilia*: IHCAS12212120.

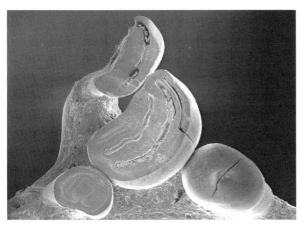

Fig. 312-11. *Cyprinus longipectoralis*: IHCAS12212139.

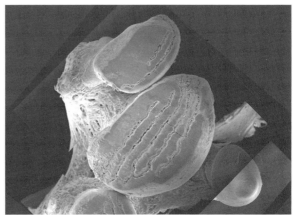

Fig. 312-12. *Cyprinus pellegrini*: IHCAS12212141.

Fig. 312-13. *Cyprinus rubrofuscus*: IHCAS122112103.

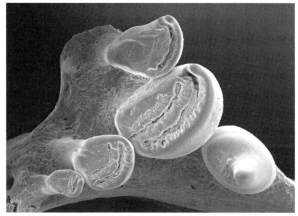

Fig. 312-14. *Cyprinus yunnanensis*: IHCAS1210016524.

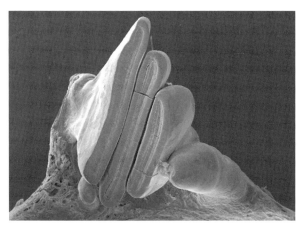

Fig. 312-15. *Carassioides acuminatus*: IHCAS12212160.

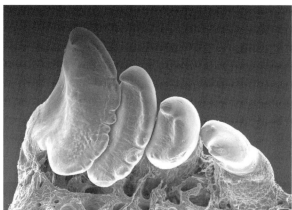

Fig. 312-16. *Carassius carassius*: LBM1210013782.

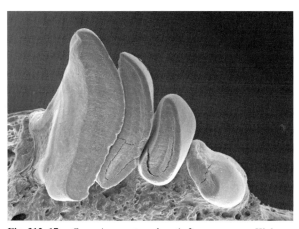

Fig. 312-17. *Carassius auratus* subsp. A. Japanese name: Kinbuna: LBM 1210042630.

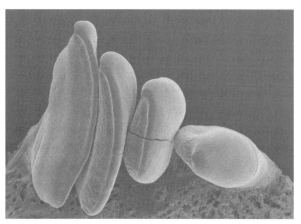

Fig. 312-18. *Carassius auratus buergeri*: NSM-P61704-5.

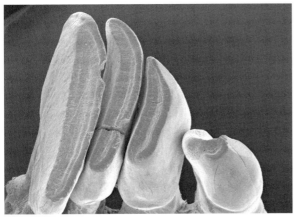

Fig. 312-19. *Carassius auratus* subsp. B. Japanese name: Nagabuna: NSM-PSK1350-1.

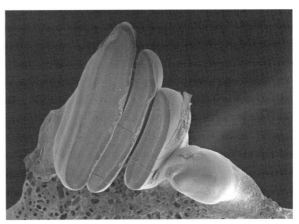

Fig. 312-20. *Carassius auratus grandoculis*: LBM1210013627.

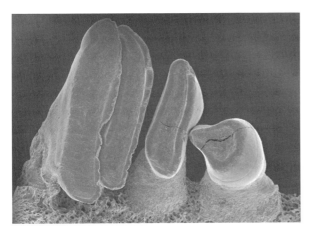

Fig. 312-21. *Carassius auratus langsdorfii*: LBM1210013630.

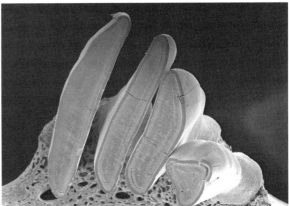

Fig. 312-22. *Carassius cuvieri*: LBM1210013620.

4. Summary

4-1. Developmental stages and types of pharyngeal teeth

My collaborators and I have examined the morphogenesis of the pharyngeal teeth of various cyprinids and shown that they go through several stages of development while becoming specialized. For example, the central teeth of cyprinines such as *Carassius auratus grandoculis* and *Cyrinus carpio* undergo the following five stages of morphogenesis:

Stage 1: Tooth recurved and conical with no grinding surface.

Stage 2: Tooth with a posteriorly pointing or slightly laterally pointing grinding surface.

Stage 3: Tooth with a laterally pointing grinding surface.

Stage 4: Tooth with a laterally pointing grinding surface and an unclear posterior margin; grinding surface either expanded and spatulate, or integrated with the posterior side of the tooth crown.

Stage 5: Tooth axis and grinding surface at approximately right angles.

The types of pharyngeal teeth that are recognized herein correspond in general to these developmental stages. Although the various stages of tooth morphogenesis are specific to species and teeth, the tooth hook, which is homologous to the tip of the Stage-1 conical tooth, may commonly be found in all stages and types of teeth. However, in teeth that have undergone specialization such as those of *Cyprinus, Mylopharyngodon,* or *Hypophthalmichthys*, only the vestiges of the tooth hook remain, and it has disappeared in some cases. Furthermore, in teeth that undergo marked wear, such as those of labeonines, xenocypridines, acheilogathines, and the cyprinine *Carassius*, the hook might have worn away completely (Fig. 4-1).

The morphological feature that is most useful as a standard for homology is the grinding surface, which is common to teeth from Stage or Type 2 to 5. The grinding surface is bounded by the tooth hook, the two occulusal margins, and the protruding shoulder of the crown. Although the grinding surface usually tends to be concave, this is not always the case (Fig. 4-2C). Because the grinding surface points posteriorly in Stage or Type 2, the two margins are at first lateral and medial and the tooth hook is at the anterior end of the crown, positioned centrally medio-laterally. In *Acrossocheilus parallens,* a barbine, all teeth from the larval stage to the adult fish only develop to Stage 1 or 2. In barbines in general, as well as in schizothracines and many danionines, tooth development stops at Stage 2. In the central teeth of barbines and schizothracines, the grinding surface becomes extended and spoon-like, and both the medial and lateral margins of the grinding surfaces as well as the tooth hooks are not only distended, but also connected to form a marginal ridge. A bulge or longitudinal ridges are observed in the center of the grinding surface in some species of barbines and schizothracines, with a U-shaped groove observed between the marginal ridge and the bulge or longitudinal ridges (Fig. 4-2).

From Stage or Type 3, the grinding surface shifts from facing posteriorly to facing laterally. At the same time, the lateral margin becomes the anterior one, and the medial one becomes the posterior one (Fig. 4-3). In such teeth, the tooth hook is at the medial end of the crown. The central teeth reach Stage 3 in subfamilies other than the Barbinae and Schizothoracinae, but only in some danionines. Even if the central teeth advance to Stage or Type 3, the anterior teeth usually remain in Stage 2, but all major-row teeth reach Stage 3 in the Hypophthalmichthyinae, Xenocypridinae, and part of the Acheilognathinae. In *Hypophthalmichthys nobilis*, the central teeth, which are positioned at A2 (An1), A3 (Ce0), and A4 (Po1), rapidly reach Stage 3 whereas the anterior teeth positioned at A1 (An2) require a long period of time to reach Stage 3 and remain at Stage 2 during the juvenile period.

Stage- or Type-4 teeth with an unclear posterior margin are often observed as central teeth in labeonines and some leuciscines, gobionines, and cyprinines. In such fish, the anterior teeth are often at Stage or Type 2. The central teeth of the leuciscine *Tribolodon*

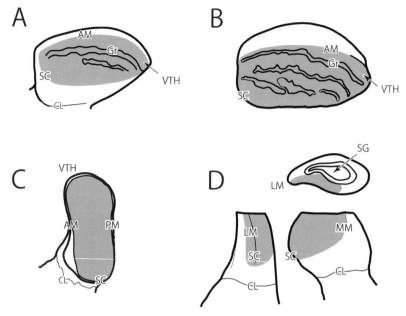

Fig. 4-1. Teeth with obscured doreroded tooth hooks. A, left tooth A3 of *Cyprinus carpio* (Cyprininae); B, left tooth A2 of *Cyprinus carpio*; C, left tooth A2 of *Hypophthalmichthys molitrix* (Hypophthalmichthyinae); D, left tooth A1 of *Carassius auratus* subsp. A (Japanese name Kinbuna) (Cyprininae). Abbreviations: AM, anterior margin of grinding surface; CL, cervical line; Gr, groove of grinding surface; MM, medial margin of grinding surface; LM, lateral margin of grinding surface; PM, posterior margin of grinding surface; SC, shoulder of tooth crown; SG, secondary grinding surface; VTH, vestige of tooth hook.

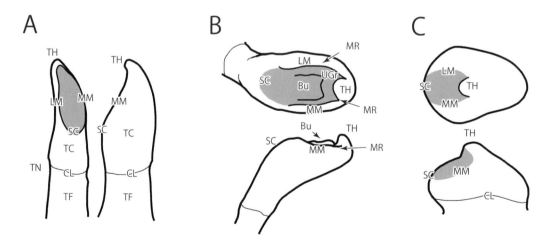

Fig. 4-2. Type-2 teeth. A, left tooth A3 of *Aaptosyax grypus* (Danioninae); B, left tooth A5 of *Acrossocheilus paradoxus* (Barbinae); C, left tooth A1 of *Cyprinus carpio* (Cyprininae). Abbreviations: Bu, bulge; CL, cervical line; LM, lateral margin of grinding surface; MM, medial margin of grinding surface; MR, marginal ridge; SC, shoulder of tooth crown; TC, tooth crown; TF, tooth foot; TH, tooth hook; TN, tooth neck; UGr, U-shaped groove.

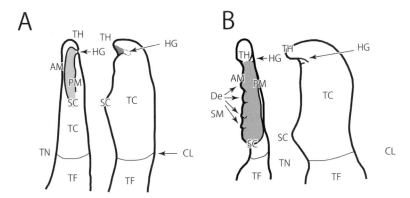

Fig. 4-3. Type-3 teeth. A, left tooth A4 of *Ischikauia steenackeri* (Cultrinae); B, right tooth A4 of *Acheilognathus rhombeus* (Acheilognathinae). Abbreviations: AM, anterior margin of grinding surface; De, denticle; HG, hamulus groove; PM, posterior margin of grinding surface; SM, serrated margin; TC, tooth crown; TF, tooth foot; TH, tooth hook; TN, tooth neck.

hakonensis, the labeonine *Cirrhinus molitorella*, and the gobionine *Gnathopogon elongatus* reach Stage 4 after developing through Stages 2 and 3. Because the teeth of *Cirrhinus molitorella* are subject to severe wear, a secondary grinding surface is formed and the tooth hooks and original grinding surface disappear. In the central teeth of *Gnathopogon elongatus*, the anterior margin of the grinding surface is curved forward, and the grinding surface becomes spatulate. Central teeth of gobionines also have a spatulate grinding surface, or else the grinding surface becomes integrated with the posterior side of the tooth crown (Type 4-1) (Fig. 4-4).

In Stage or Type 5, the tooth axis and the grinding surface are orientated at approximately right angles to each other (Fig. 4-5). The central teeth of *Mylopharyngodon piceus*, *Cyprinus carpio*, and *Carassius auratus grandoculis* develop through the first four stages before reaching Stage 5. In larvae or juveniles of *Mylopharyngodon*, *Cyprinus*, and *Carassius*, the Stage-4 grinding surface of these teeth expands into a sort of spatula in which a shallow groove can be observed (Type 4-1). In *Cyprinus carpio*, this groove becomes narrow, deep, and clearly visible, and it is typical in *Cyprinus*, for the numbers of grooves of the grinding surface to increase over the course of development. In *Cyprinus carpio*, although tooth A2 (An1) reaches Stage 5, tooth A3 (Ce0) only reaches Stage 4, and tooth A1 (An2) stops at Stage 2. In full grown *Cyprinus carpio*, tooth A3 also reach stage 5, whereas the A1 teeth are depressed, and the tooth hooks of all teeth become vestiges or are lost. In *Carassius auratus grandoculis*, although teeth A2 (An1), A3 (Ce0), and A4 (Po1) reach stage 5, tooth A1 (An2) stop in stage 2.

4-2. Dentition of each subfamily

In danionines, teeth are generally arranged in three rows except except for *Pararasbora*, *Aphyocypris*, *Parazacco*, and *Gobiocypris*, all of which bear two rows of teeth, and *Esomus* with one row. Although the number of teeth is rich in variation, the most typical dental formula is 2.4.5-5.4.2 in species with three-row dentitions. Ridges and bulges hardly develop on the grinding surfaces of most species. Danionine central teeth are Type 2 (Type 2-1, 2-2, or 2-3) except for *Esomus* (Type 3), *Aspidoparia* (Type 5), and *Gobiocypris* (Type 4). Fish bearing various dentitions are included in this subfamily, and danionines seem not to be monophyletic.

In barbines, the teeth are always arranged in three rows. Although the number of teeth varies greatly, the most typical dental formula is 2.3.5-5.3.2 and all the teeth are Type 2 except in *Catlocarpio* with Type 4 or 5 teeth in its one-row dentition. The central teeth are usually Type 2-2, but Type 2-1 in *Hampala*, *Percocypris*, and some species of *Acrossocheilus*, *Anematichthys*, *Balantiocheilos*, *Barbodes*, and *Barbus*. In contrast to danionines, no central teeth are of Type 2-3. Bulges, ridges, or grooves are often developed on the grinding surface of Type 2-2 teeth. Barbines except for *Catlocarpio* have uniform dentitions and seem to be a monophyletic.

In schizothoracines, the teeth are arranged in three

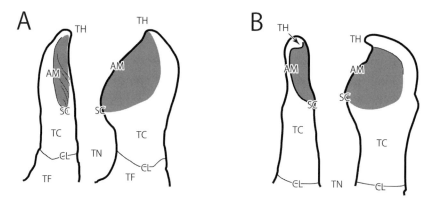

Fig. 4-4. Type-4 teeth. A, left tooth A4 of *Hemibarbus labeo* (Gobioninae); B, left tooth A4 of *Pseudogobio esocinus* (Gobioninae). Abbreviations: AM, anterior margin of grinding surface; CL, cervical line; SC, shoulder of tooth crown; TC, tooth crown; TF, tooth foot; TH, tooth hook; TN, tooth neck.

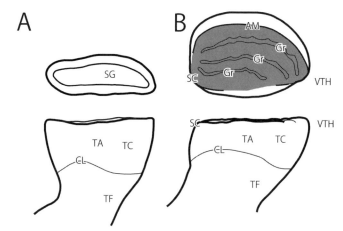

Fig. 4-5. Type-5 teeth. A, left tooth A2 of *Carassius aruratus langsdorfi* (Cyprininae); B left tooth A2 of *Cyprinus carpio* (Cyprininae). Abbreviations: AM, anterior margin of grinding surface; CL, cervical line; SC, shoulder of tooth crown; SG, secondary grinding surface; TA, tooth axis; TC, tooth crown; TF, tooth foot; VTH, vestige of tooth hook.

or two rows. The number of teeth varies little. The typical dental formula is 2.3.5-5.3.2 in *Schizothorax* as in barbines, and 3.4-4.3 in *Gymnocypris*, *Gymnodiptychus*, *Ptychobarbus*, and *Schizopygopsis*. All the teeth are Type 2: the central teeth and minor-row teeth are always Type 2-2 and the anterior teeth are usually Type 2-1. No central teeth are Type 2-3 as in barbines. Bulges, ridges, or grooves are usually developed on the grinding surfaces of Type 2-2 teeth. Schizothoracines have a uniform dentition and seem to be monophyletic.

In cultrines, the teeth are arranged in three rows. Although the variation in number of teeth is rich, the most typical dental formula is 2.4.5-5.4.2, as in danionines. The central teeth are usually Type 3, the anterior teeth Type 2, and the minor-row teeth Type 2 or 3. A hamulus groove is present on the posterior or medial side of the crown in the central teeth and minor-row teeth in almost every species. Cultrines have a uniform dentition and seem to be monophylrtic.

In xenocypridines, the teeth are arranged in one to three rows. The number of teeth in the major row is more than five, usually six. The typical dental formula is 2.4.6-6.4.2 in *Xenocypris* and *Plagiognathops*, 3.6-6.3 in *Distoechodon*, and 6-6 in *Psudobrama*. All the teeth are

Type 3 (Type 3-0) and intensely worn with a secondary grinding surface. The major-row teeth are compressed antero-posteriorly, and the minor-row teeth are slender and rod-like. Xenocypridines have a uniform dentition and seem to be monophyletic.

The dental formula is 4-4 in hypophthalmichthyines, and no variation of tooth number occurs. All teeth are Type 3 (Type 3-1), depressed and slipper-shaped. Hypophthalmichthyines have a uniform dentition and seem to be monophyletic.

Acheilognathine teeth are arranged in one row, and the dental formula is always 5-5 without exception. All teeth are Type 3-0 and compressed antero-posteriorly. A hamulus groove is present on the posterior side of the crown of major-row teeth other than tooth A1 (An3). The tooth crowns are worn and secondary grinding surfaces are frequently formed. Acheilognathines have a uniform dentition and seem to be monophyletic.

Leuciscine teeth are arranged in one or two rows except for *Elopichthys*, *Ochetobius*, and *Squaliobarbus* which have three-row dentitions. The number of teeth varies but little, and the most typical dentition is 2.4.5-5.4.2 in species with three-row dentitions, as in Danioninae and Cultrinae, 2.5-4.2 in species with two-row dentitions, and 4-4 in species with one-row dentitions. In *Alburnoides* with its single-row dentition, dental formula is 6 -5, and the number of major-row teeth is more than 5 as in Xenocypridinae. Although the central teeth of most species of Leuciscinae are Type 3 or 4, Type-2 central teeth occur in *Elopichthys*, and Type 5 in *Tinca* and *Mylopharyngodon*. Fish bearing various dentitions are included in this subfamily, and leuciscines seem not to be monophyletic.

Labeonine dentitions are uniform: their teeth are always arranged in three rows. There is little variation in tooth number, and the most typical dental formula is 2.4.5-5.4.2 as in Danioninae and Cultrinae. The central teeth are generally Type 4. Most of labeonine teeth are well worn, and secondary grinding surfaces are frequently formed. A hamulus groove is usually present on the posterior side of major-row teeth. Labeonines seem to be monophyletic, judging from their dental characteristics.

Gobiobotine teeth are arranged in two rows. the most typical dental formula is 3.5-5.3. Variation in tooth number is rare. The central teeth are Type 4 (Type 4-1), and the anterior teeth and minor-row teeth are Type 2-1 in all species. Gobiobotines seem to be monophyletic,

judging from their dental characteristics.

In gobionines, the teeth are arranged in one or two rows except for *Beligobio* and *Hemibarbus*, which have three-row dentitions. The dental formula is 1.3.5-5.3.1 almost without exception in species with three-row dentitions, 3.5-5.3 in species with two-row dentitions but with some variation in tooth number, and 5 - 5 without exception in species with single-row dentitions. The central teeth are Type 4 (Type 4-1) in all species. Gobionines seem to be monophyletic based on their dental characteristics.

In cyprinines, the central teeth in *Puntioplithes* and *Procypris* are Type 2 (Type 2-2), and their dental formula is 2.3.4(5)-4(5).3.2; their dentition is thus like that of barbines. Although the central teeth are generally Type 5 in *Cyprinus*, *Carassioides*, and *Carassius*, the most posterior central tooth (tooth A3: Ce0) is Type 4-1 in some species of *Cyprinus*. The teeth are arranged in three rows in *Cyprinus*, two rows in *Carassioides*, and one row in *Carassius*, with the dental formulae 1.1.3-3.1.1, 2.4-4.2, and 4-4, respectively. Variation in tooth number hardly arises. Cyprinines except for *Puntioplithes* and *Procypris*, i.e., those with Type-5 central teeth, seem to be monophyletc.

According to morphology of the pharyngeal dentitions, schizothoracines, cultrines, xenocypridines, hypophthalmichthyines, acheilognathines, labeonines, gobiobotines, and gobionines each appear monophyletic, but in other subfamilies, certain species do not match their taxa. *Esomus*, *Aspidoparia*, and *Gobiocypris* should be excluded from the Danioninae. The dentition of *Gobiocypris* corresponds to that of the Gobioninae, with a dental formula of 2.4-4.2 and Type-4 central teeth. The dentitions of *Catlocarpio* are quite different from those of most barbines. *Elopichthys* and *Mylopharyngodon* should be excluded from the Leuciscinae, and *Elopichthys* dentitions are of those of the Danioninae. Finnaly, the dentitions of *Puntioplithes* and *Procypris* match those of barbines, not those of typical cyprinines.

4-3. Nomenclature of pharyneal dentition

Rutte (1962) summarized the nomenclature of the parts of pharyngeal teeth, but his list of terms is insufficient for describing all the various forms. Based on subsequent morphogenetic studies, the homologies of various forms of pharyngeal teeth are shown in Fig. 4-1 to 4-5 using the relative positions of the tooth hook, the

4. Summary   151

shoulder of the crown, and the grinding surface bounded by its two occlusal margins as markers. Based on the homologies indicated, the nomenclature of the individual parts of the pharyngeal tooth was redefined.

The pharyngeal tooth consists of a bulging crown and a cylindrical tooth foot joined to the pharyngeal bone, with the narrow tooth neck located between them. The crown portion is exposed to the pharyngeal cavity and is glossy due to being covered with an enameloid layer. The cervical line is the border between the crown and foot (Fig. 4-1 to 4-5).

With the crown located at the apex of the tooth, its four sides consisting of the medial, lateral, anterior, and posterior surfaces are observed. During Stage 2, the shoulder of the crown protrudes from the posterior surface, with the grinding surface formed between the tooth hook and the crown's shoulder. The orientation of the grinding surface becomes altered at Stage 3, when it moves to the lateral side. Generally, the crown tends to be compressed antero-posteriorly at Stage 3 and 4, resulting in many teeth with narrow medial and lateral surface and wide posterior and anterior surfaces.

The occlusal margins have small projections and denticles lined up to form serrations. The serrate margin is observed in the anterior margin of acheilognathine teeth. In *Ctenopharyngodon idella*, grooves and ridges in a comb-like pattern can be observed on the anterior and posterior margins and surfaces of the crown. These are also derived from marginal denticles on margins and homologous to serrate margins. In addition, at the base of the tooth hook, groove-like indentations, termed the hamulus groove, can be observed extending from the grinding surface to the crown's posterior surface. The hamulus groove is often observed in cutrines, acheilognathines, and labeonines (Fig. 4-3).

Morphogenesis of tooth types differs in various fish species depending on the stage. As described above, the criteria determining the stage of each tooth are the tooth hook, its traces, and the shoulder of crown position. Stage 2 is characterized by the tooth hook being at anterior end of the crown, whereas stage 3 or later is defined by the tooth hook being at medial end of the crown. Comparing the anterior with the lateral margin and the posterior with the medial margin, both are identical during the early stages of morphogenesis (Stage 2 to 3). However, as the stages progress (Stages 4 and 5), the anterior margin clearly curves forward in a convex shape whereas the posterior margin becomes ambiguous.

Terminology for pharyngeal dentitions

adult dentition: Commonly represented by a dental formula in systematic descriptions, the adult dentition is formed by the early stage of the juvenile period and is seen in both juveniles and adults. The major and minor tooth rows have all been formed, and a tooth row consists of teeth formed in multiple replacement waves.

anterior position: Tooth positions before position An1, *viz,* positions An2, An3, and so on, in the major row. In the cyprinid larval dentition and the major row of the adult dentition, they are the positions before the three rearmost positions.

anterior tooth: Teeth appearing at anterior postions An2, An3, and so on. In the adult dentition of cyprinids, they are the teeth before the three rearmost positions in the major row. Anterior teeth do not appear in cobitid and catostomid dentitions.

central position: The three positions Po1, Ce0, and An1. They are the three rearmost positions in the cyprinid larval dention and the major row of the adult dentition, and the three foremost positions in dentitions of cobitids or catostomids.

central tooth: Teeth appearing at the central positions Po1, Ce0, and An1.

family progenitor: The initial tooth of each tooth family. Tooth $_1$[An1] is the family progenitor of tooth family An1.

larval dentition: A multi-row dentition generally seen in larvae, in which the teeth of each row belong to a single replacement wave. Each tooth row in the larval dentition thus consists of teeth ankylosed alternately at even and odd positions.

major row: The most medial tooth row (A row) of the adult dentition. The larval dentition develops into this row.

Major-row tooth: Teeth of the major row, including the central teeth and anterior teeth.

minor row: Lateral tooth rows (B and C rows) of the adult dentition. The minor rows represent a different system from the larval dentition or the major row of the adult dentition.

Minor-row tooth: Teeth of the minor rows.

perfect type of larval dentition: A larval dentition in which the first few teeth at anterior positions appear without fail as in A-type and C-type larval dentitions.

posterior position: Tooth positions behind position Po1, *viz*, positions Po2, Po3, and so on, in cobitid and catostomid dentitions.

posterior tooth: Teeth apparing at posterior positions Po2, Po3 and so on. They are never seen in cyprinid dentitions.

replacement wave: A set of replacement teeth of the same generation, which appear at alternate positions (even or odd).

stem progenitor: The initial tooth of the dentition. Tooth $_0$[Ce0] is the stem progenitor of the larval dentition and of the major row of the adult dentition.

tooth family: The intial tooth (family progenitor) and the replacement teeth that in sequence appear at each tooth position.

tooth name: In the adult dentition, tooth position serves as the tooth name. For example, tooth A1 is the all-inclusive term for the teeth that successively appear at position A1 and belong to the same tooth family. In the larval dentition, The general formula for position (r) and replacement wave number (n) is $_{n-1}$[r]; thus the tooth position Ce0 in the first replacement wave, which is where the initial tooth (I) arises, would be identified as $_0$[Ce0].

tooth position: In the adult dentition, the tooth positions are numbered from anterior to posterior in each row. In the larval dentition, the tooth position where the intial tooth appears is numbered Ce0; more anterior tooth positions are numbered successively as An1, An2, An3, An4, An5, etc., and more posterior ones are numbered successively as Po1, Po2, etc. If there are five major-row teeth, position A1 of the adult dentition corresponds to position An3 of the larval dentition. If there are four major row teeth, position A1 corresponds to position An2.

tooth row: Teeth that form a line along the bone. In the adult dentition, each tooth row consists of teeth belonging to multiple replacement waves. In the larval dentition, each tooth row consist of teeth of a single replacement wave.

transitional dentition: During the period between the appearance of the first minor-row tooth and the completion of the adult dentition, a "transitional dentition" is present. It can be difficult to identify each tooth in the lateral part of the pharyngeal bone in the transitional dentition.

imperfect type of larval dentition: A larval dentition in which the first few teeth at anterior positions do not appear, but only their later replacements, as in B-type and D-type larval dentitions.

Terminology for tooth parts

angle of grinding surface: The angle between the tooth axis and the grinding surface.

anterior margin: The front edge of the grinding surface, which was originally the lateral margin.

bulge: An elevation on the grinding surface seen in some barbine and schizothoracine teeth.

cervical line: An enamel-bone junction, the dividing line between the crown and foot portions of a tooth.

crown (tooth crown): The part of the tooth covered with an enameloid layer and exposed to the pharyngeal cavity.

denticle: A small, tooth-like prtojection observed on the grinding surface and/or the margins of the grinding surface in some fish.

grinding surface: A surface bounded by the lateral (anterior) and medial (posterior) margins, the shoulder of the crown, and the hook. In *Cyprinus*, the dorsal surface of the crown of tooth A2 is generally called a grinding surface, but it is not a true grinding surface; the latter is represented by the most anterior groove on the dorsal surface.

groove: A slot-like hollow on the functional grinding surface that is a remnant of the original concave grinding surface.

hamulus groove: An incisure-like slot leading to a concave grinding surface, located at the base of the tooth hook on the posterior side of the tooth crown.

lateral margin: The outside edge of the grinding surface, which is homologous to the anterior margin.

marginal ridge: A continuous ridge with both the medial and lateral margins of the grinding surface as well as the tooth hook swollen.

margins (of the grinding surface): The two edges of the grinding surface, being fundamentally medial and lateral in Type-2 teeth, but becoming posterior and anterior in Type-3 to -5 teeth.

medial margin: The inside edge of the grinding surface, which is homologous to the posterior margin.

ridge: A linear elevation on the grinding surface seen in some barbine and schizothoracine teeth.

posterior margin: The trailing edge of the grinding surface, which was originally the medial margin.

secondary grinding surface: Wear mark produced on a tooth crown when a tooth is strongly worn and dentin is exposed.

serrate margin: Saw tooth-like row of denticles on the margins.

shoulder (of crown): A projection in the middle of the

tooth crown with the grinding surface formed between the shoulder and the tooth hook.

tooth axis: The central axis of the tooth crown.

tooth crown: The upper portion of the tooth distal to the cervical line, which is exposed to the pharyngeal cavity.

tooth foot: The cylindrical proximal part of a tooth, which combines with connective tissue and ankyloses to the pharyngeal bone at its base.

tooth hook: A hooked projection at the tip of a tooth. Almost all teeth bear a hook or a vestige of a hook, except for very worn teeth.

tooth neck: The narrow part of a tooth in the vicinity of the cervical line.

U-shaped groove: A groove between the marginal ridge and the bulge on the grinding surface in some barbines and schizothoracines.

References

Berg, L. S., 1940, Classification of fishes, both recent and fossil. *Trav. L'Inst. Zool. L'Acad. Sci. L'URSS*, 5:87‒517.

Chu, Y.-T., 1935, Comparative studies on the scales and on the pharyngeal and their teeth in Chinese cyprinids, with particular reference to taxonomy and evolution. *Biol. Bull. St. Johns Univ. 2.*

Chen, Y.-Y. *et al.*, 1998, *Fauna Sinica, Osteichthyes, Cypriniformes II*. Science Press, Beijing. (in Chinese)

DeMar, R., 1972, Evolutionary implication of Zahnreihen. *Evolution*, 26: 435‒450.

Eastman, J. T. and Underhill, J. C., 1973, Intraspecific variation in the pharyngeal tooth formulae of some cyprinid fishes. *Copeia*, 1973(1): 45‒53.

Edmund, A. G., 1960, Tooth replacement phenomenon in the lower vertebrates. *Contr. Life Sci., Dv. Roy. Ont. Mus.*, 52: 1‒190.

Edmund, A. G., 1962, Sequence and rate of tooth replacement in the Crocodilia. *Contr. Life Sci., Dv. Roy. Ont. Mus.*, 56: 1‒42.

Evans, H. E. and Deubler, E. E. Jr., 1955, Pharyngeal tooth replacement in *Semotilus stromaculatus* and *Clinostomus elongates*, two species of cyprinid fishes. *Copeia*, 1955: 31‒34.

He, S.-P., Yue, P.-Q., and Chen, Y.-Y., 1994, The development of the pharyngeal dentition in a cyprinid, *Gobiocypris rarus* Fu et Ye. *Acta Hydrobiologica Sinica*, 18(2): 150‒155. (in Chinese)

He, S.-P., Yue, P.-Q., and Chen, Y.-Y., 1997, Comparative study on the morphology an Development of the pharyngeal dentition in the families of Cypriniformes. *Acta Zoologica Sinica*, 43(3): 255‒262. (in Chinese)

Husysseune, A., Van der Heyden, C., and Sire, J.-Y., 1998, Early development of the zebrafish (*Danio rerio*) pharyngeal dentition (Teleostei, Cyprinidae). *Anat. Embryol.* 198: 289‒305.

Huysseune, A., Van der heyden, C., and Sir, J.-Y., 2000, Early development of the zebrafish (*Danio rerio*) pharyngeal dentition (Teleostei, Cyprinidae). *Anat. Embryol*, 198:289‒305.

Kodera, H., 1982, Morphodifferentiation of pharyngeal teeth of the carp, *Cyprinus carpio*. *Tsurumi Univ. dent. J.*, 8(2): 179‒212. (in Japanese)

Lawson, R., Wake, D. B., and Beck, N. T., 1971, Tooth replacement in the red-backed salamander, *Plethodon cinereus*. *J. Morphol.* 134: 259‒269.

Nakajima, T., 1979, The development and replacement pattern of the pharyngeal dentition in the Japanese cyprinid fish, *Gnathopogon carrulescens*. *Copeia*, 1979(1): 22‒28.

Nakajima, T., 1984, Larval vs. adult pharyngeal dentition isn some Japanese cyprinid fishes. *J. Dent. Res.* 63(9): 1140‒1146.

Nakajima, T., 1987, Development of pharyngeal dentition in the cobitid fishes, *Misgurnus anguillicaudatus* and *Cobitis biwae*, with consideration of evolution of cypriniform dentions. *Copeia*, 1987(1): 208‒213.

Nakajima, T., 1990, Mophogensis of the pharyngeal teeth in the Japanese dace, *Tribolodon hakonensis* (Pisces: Cyprinidae). *J. Morphol.*, 205: 155‒163.

Nakajima, T., 1994, Succession of cyprinid fauna in Paleo-lake Biwa. *Arch. Hydrobiol. Neih. Ergebn. Limnol.*, 44: 433‒439.

Nakajima, T., 1998, Koi-ka gyorui intoushi no keitai bunka to aka no bunrui (Morphogenesis of cyprinid pharyngeal teeth and classification of cyprinid subfamilies). *Urban Kubota* (37): 36‒37. (in Japanese)

Nakajima, T., 2006, Significance of freshwater fisheries during the Jomon and Yayoi periods in western Japan: An analysis of pharyngeal tooth remains of cyprinid fishes. Grier, C., Kim, J., and Uchiyama, J. eds. "*Beyond Affluent Foragers*" pp. 45‒53, Oxbow Books, Oxford.

Nakajima, T., 2012, Origin and Temporal Succession of the Cyprinid Fish Fauna in Lake Biwa. Kawababe, H., Nishino, M. and Maehata, M. eds. "*Lake Biwa: Interactions between Nature and People*" pp.17‒23, Springer.

Nakajima, T., Nakajima, M., and Yamazaki, T., 2010, Evidence for fish cultivation during the Yayoi period in western Japan. *Int. J. Osteoarchaeol.* 20: 127‒134.

Nakajima, T., Sugito, M., Nakahara, M., and Ozaki, M., 1981, An analysis on the pattern of tooth replacement in the cyprinid fish, *Rhodeus ocellatus ocellatus. Jpn. J. Oral Biol.*, 23: 893–895.

Nakajima, T. and Yamasaki, H., 1992, Temporal and spatial distribution of fossil cyprinids in east Asia and their paleogeographic significance. *Bull. Mizunami Fossil Mus.*, 19: 542–557. (in Japanese)

Nakajima, T., Yoshida, H., Sone, B., and Hotta, Y., 1983, Replacemrnt pattern of the pharyngeal teeth in cyprinid fish, *Tribolodon hakonensis. Jpn. J. Oral Biol.*, 25: 801–803.

Nakajima, T. and Yue, P.-Q., 1989, Development of the pharyngeal teeth in the big head, *Aristichthys nobilis* (Cyprinidae). *Jpn. J. Ichthyol.*, 36(1): 42–47.

Nakajima, T. and Yue, P.-Q., 1995, Morphological changes in development of pharyngeal teeth in *Mylopharyngodon piceus. Chin. J. Oceanol. Limnol.*, 13(3): 271–277.

Nelson, M. V. H., 2016, *Fishes of the World Fifth Edition*. John Wiley and Sons, Hoboken.

Osborn, J.W., 1971, The ontogeny of tooth succession in *Lacerta vivipara* Jacquin (1987). *Proc. R. Soc. Lond. Ser. B*, 179: 261–289.

Osborn, J.W., 1972, On the biological improbability of Zhanreihen as embryological units. *Evolution*, 26: 601–607.

Osborn, J.W., 1974, On the control of tooth replacement in reptiles and its relationship to growth. *J. Theor. Biol.*, 46: 509–527.

Osborn, J.W., 1978, Mophogentic gradients: Fields versus clones. Butler, P. M. and Joysey, K. A. eds. *"Development, Function and Evolution of teeth"*, pp. 171–201, Academic Press, New York.

Regan, C. T., 1912, The anatomy and classification of the teleostean fishes of the Order Lyomeri. *Ann. Mag. Nat. Hist.* Ser. 8, 10: 13–32.

Rutte, E., 1962, Schlundzhane von Susswasserfisschen. *Paleontographica* Abt. A, 120: 165–212.

Sato, T., Kido, Y., Hamaguchi, H., and Nakajima, T., 2000, Morphological defferenciation of the pharyngeal teeth in Gnathopogon elongatus. *Jpn. J. Ichthyol.*, 47(2): 109–114. (in Japanes)

Shan, X.-H., 2001, Development and replacement of pharyngeal teeth in the cyprinid fish, *Distoechodon compressus. Acta Hydrobiologia Sinica*, 25(1): 42–49. (in Chinese)

Shkil, F. N., Levin, B. A., Abdissa, B., and Smirnov, S. V.,2010, Variability in the number of tooth rows in the pharyngeal dentition of *Barbus interdedius* (Teleostei; Cyprinidae): genetic, hormonal and environmental factors. *J. Appl. Ichthyol.*, 26: 315–319.

Yue P.-Q. and Nakajima, T., 1994, Development of the pharyngeal teeth in Chinese labeonine fish *Cirrhinus molitorella. Zool Res.*, 15: 74–81.

Van der Heyden, C. and Huysseune, A., 2000, Dynamics of teeth formation and replacement in the zebrafish (*Danio rerio*) (Teleostei, Cyprinidae). *Developmental Dynamics*, 219: 486–496.

Van der Heyden, C., Wautier, K., and Huysseune, A., 2001, Tooth succession in the zebrafish (*Danio rerio*). *Arch. Oral Biol.*, 46: 1051–1058.

Vasnecov, V. V., 1939, Evolution of the pharyngeal teeth in Cyprinidae. *A la memoire de A. N. Severtzoff*, 16: 439–491. (in Russian)

Vladykov, V. D., 1934, Geographical variation in the nuber of rows of pharyngeal teeth in cyprinid genera. *Copeia*, 1934(3): 134–136

Weisel, G. F., 1967, The pharyngeal teeth of larval and juvenile suckers (*Catostomus*). *Copeia* 1967(1): 50–54.

Postscript

This volume comprises images of pharyngeal teeth taken over the past 35 years. For that reason, I ask your forgiveness for the varying levels of image quality. I began collecting specimens of cyprinid pharyngeal teeth during my student days. I personally amassed approximately 2,500 specimens, which are preserved in the Lake Biwa Museum. In addition, specimens of freshwater fish from all over China are preserved in the Institute of Hydrobiology, Chinese Academy of Sciences. I visited this institute in Wuhan for the first time in 1981 and was astounded by the collection, and I have visited countless times since then. The images in this volume are taken from both my specimens and those of the Institute of Hydrobiology.

The purpose of this volume is to show how the pharyngeal dentition in various groups (subfamilies) of Cyprinidae differ by group, not to describe the differences in teeth by species. A separate publication to aid identification at the species level is planned. The pharyngeal teeth of the species of Cyprinidae are an extremely valuable key for the classification of the species of this family. Here I show how the pharyngeal teeth of certain species of cyprinids whose shape varies by group were formed ontogenetically from similar simple conical teeth. The teeth change their shape as they are replaced, passing through several stages from the conical stage. I show that the stage of development reached by different teeth is different for each group.

Photographing early-stage dentition with a scanning electron microscope (SEM) is not easy. The hardships involved in preparing samples for SEM are many. For any given species, the first teeth observed are from newly-hatched larvae of 5 mm in body length. Pharyngeal teeth could be extracted from such small fish under a stereomicroscope, but only after fish fixed in formalin had been dipped in an aqueous solution of potassium hydroxide and alizarin red S and their pharyngeal bones stained red. The pharyngeal bone was extracted along with the soft tissue surrounding it using a tungsten needle and titanium tweezers. The resected pharyngeal bone is small enough to lose sight of easily. To create a clean sample, the soft tissues and the mucous membranes around the pharyngeal teeth and bones must be removed. This required me to maximize the magnification of the stereomicroscope. In the microscopic world, sharpened titanium tweezers are as soft as paper, so the soft tissue between the teeth has to be removed with a tungsten needle. One must remove the soft tissue surrounding the teeth with the needle without moving one's hands, or else the sample moves out of the field of view.

I wrote this volume with my memories of these tales of struggle and my annual visits to Wuhan, where I holed myself up in the specimen vault while also watching China change these past 35 years.

On a final note, I am indebted to many people in writing this volume. I received a great deal of assistance from late Dr. Yoshio Tomoda and the late Dr. Takeichiro Kafuku, who first gave me the opportunity to study the pharyngeal teeth of cryprinid fish. Professors Zhen Yiyu, Yue Peiqi, He Shunping, and Liu Huangzhan and the rest of the staff at the Institute of Hydrobiology, Chinese Academy of Sciences, accommodated me so that I could use the samples at the Institute. I received technical advice on taking SEM images from Dr. Haruto Kodera of Tsurumi University. In addition, many people assisted me in the collection of fish specimens. Mr. Katsuyuki Fujimoto, Mr. Tomoyuki Sato, and the members of the "Uonokai (Fish Survey Club)" collected specimens. In particular, Mr. Tomoyuki Sato collected a number of Cambodian species of cyprinid fish. Dr. Mark J. Grygier of the Lake Biwa Museum and, later, National Taiwan Ocean University was my English editor. Mr. M. Matsuda of the Lake Biwa Museum went through the procedure to lend me the pharyngeal specimens of the Lake Biwa Museum. Mr. Daisuke Hirota, graduate student, and other students at the Paleoichthyological Laboratory of the Okayama University of Science, helped me to put the data in order. I am greatly indebted to Mr. Hidefumi Ina of Tokai University Press for his editing of this volume. I would like to express my heartfelt thanks to these people. A part of this work was supported by JSPS KAKENHI Grant Numbers JP22401002, JP26300004.

Finally, my wife Michiyo, in addition to helping me organize the data and take photographs, was an unflagging source of support in the publication of this volume. Without this support, this volume would have never seen the light of day. I cannot be too grateful.

Index

Scientific name

A

Aaptosyax grypus   29, 148

Abbotina rivularis   129

Acanthogobio guentheri   126

Acanthorhodeus chankaensis   93

Acheilognathinae   92, 147, 149

acheilognathine   147, 151, 152

Acheilognathus cyanostigma   93

Acheilognathus intermedia   93

Acheilognathus longipinnis   93

Acheilognathus macropterus   93

Acheilognathus melanogaster   93

Acheilognathus rhombeus   18, 25, 93, 149

Acheilognathus tabira erythropterus   94

Acheilognathus tabira jordani   94

Acheilognathus tabira nakamurae   94

Acheilognathus tabira tabira   94

Acheilognathus tabira tohokuensis   94

Acheilognathus tonkinensis   94

Acheilognathus typus   94

Acheilognathus yamatsutae   94

Acrossocheilus   15, 149

Acrossocheilus beijiangensis   45

Acrossocheilus fasciatus   45

Acrossocheilus hemispinus   46

Acrossocheilus iridescens   45

Acrossocheilus kreyenbergii   45

Acrossocheilus longipinnis   45

Acrossocheilus monticola   46

Acrossocheilus paradoxus   46, 148

Acrossocheilus parallens   11, 13, 15, 25, 46, 147

Acrossocheilus wenchowensis   46

Acrossocheilus yunnanensis   46

Alburnoides   100, 151

Alburnoides bipunctatus   101

Anabarilius grahami   81

Anabarilius macrolepis   81

Anabarilius polylepis   81

Anabarilius transmontanus   81

Ancherythroculter lini   81

Anematichthys   149

Anematichthys armatus   46

Anematichthys repasson   46

Aphyocypris   29, 149

Aphyocypris chinensis   33

Aphyocypris kikuchii   33

Aspidoparia   29, 149, 151

Aspidoparia morar   34

B

Balantiocheilos   149

Balantiocheilos melanopterus   46

Barbichthys laevis   113

Barbinae   45, 139, 147, 148

barbine   147, 149-151

Barbodes   149

Barbodes semifasciolatus   47

Barbodes wynaadensis   47

Barbonymus altus   47

Barbonymus gonionotus   47

Barbus   149

Barbus afrohamiltoni   47

Barbus barbus   47

Barbus bifrenatus   47

Barbus bynni   47

Barbus jae   47

Barbus paludinosus   48

Barbus peloponnesius   48

Barbus tauricus   47

Barbus toppini   48

Barbus trimaculatus   48

Barilius bendelisis   29

Beligobio   120, 126, 151

Belligobio nummifer   126

Belligobio pengxianensis   126

Biwia zezera   130

Brevibora dorciocellata   29

C

Campostoma anomalum   101

Candidia barbata   32

Index   159

Capoeta capoeta gracilis 55

Carassioides 139, 151

Carassioides acuminatus 140

Carassius 9, 15, 139, 147, 149, 151

Carassius auratus buergeri 140

Carassius auratus grandoculis 4, 9-11, 15, 25, 141, 147, 149

Carassius auratus langsdorfii 141, 150

Carassius auratus subsp. A 140, 148

Carassius auratus subsp. B 141

Carassius carassius 140

Carassius cuvieri 141

Catla catla 113

Catlocarpio 45, 149, 151

Catlocarpio siamensis 55

catostomid 8

Chanodichthys dabryi 80

Chanodichthys erythropterus 80

Chanodichthys mongolicus 80

Chrosomus eos 101

Chrosomus erythrogaster 101

Cirrhinus microlepis 114

Cirrhinus molitorella 9, 20, 25, 114, 149

Cirrhinus mrigala 114

Clinostomus elongatus 102

cobitid 6-8

Coreius guichenoti 130

Coreius heterodon 130

Coreoleuciscus splendidus 126

Cosmochilus harmandi 48

Couesius plumbeus 102

Crossocheilus oblongus 115

Ctenopharyngodon idella 4, 20, 25, 102, 152

Culter alburnus 80

Culter oxicephaloides 80

Culter recurviceps 81

Cultrinae 80, 100, 149, 151

cultrine 150-152

Cyclocheilichthys apogon 48

Cyclocheilichthys enoplos 48

Cyclocheilichthys furcatus 48

Cyclocheilichthys lagleri 48

Cyprinella spiloptera 102

cyprinid 1, 2, 6-9

cypriniform 9

Cypriniformes 6

Cyprininae 4, 139, 148, 150

cyprinine 9, 147, 150, 151

Cyprinion kais 48

Cyprinus 139, 147, 149, 151

Cyprinus barbatus 139

Cyprinus carpio 4, 6, 9, 21, 25, 140, 147-149

Cyprinus chilia 140

Cyprinus longipectoralis 140

Cyprinus multitaeniatus 139

Cyprinus pellegrini 140

Cyprinus rubrofuscus 140

Cyprinus yilongensis 139

Cyprinus yunnanensis 140

D

Danio 29

Danioninae 4, 29, 45, 80, 100, 148, 151

danionine 147, 149

Danio albolineatus 31

Danio dangila 31

Danio erythoromicron 31

Danio kerri 31

Danio rerio 5, 32

Desmopuntius johorensis 49

Devario 29

Devario aequipinnatus 31

Devario auropurpureus 31

Devario devario 31

Devario malabaricus 31

Distoechodon 89, 150

Distoechodon hupeinensis 89

E

Elopichthys 100, 151

Elopichthys bambusa 100

Epalzeorhynchos bicolor 114

Epalzeorhynchos frenatus 114

Epalzeorhynchos munense 114

Esomus 29, 149, 151

Esomus caudiocellatus 33

Esomus danricus 34

Esomus longimanus 34

Esomus metallicus 33

F

Folifer brevifilis 49

G

Garra cambodgiensis   113

Garra gotyla gotyla   113

Garra orientalis   113

Gnathopogon   15

Gnathopogon caerulescens   1, 4, 126

Gnathopogon elongatus   4, 9, 14, 15, 25, 127, 149

Gnathopogon herzensteini   127

Gnathopogon imberbis   127

Gnathopogon nicholsi   127

Gnathopogon taeniellus   127

Gnathopogon tsinanensis   127

Gobiobotia abbreviata   123

Gobiobotia brevibarba   123

Gobiobotia filifer   123

Gobiobotia guilingensis   123

Gobiobotia kolleri   123

Gobiobotia meridionalis   123

Gobiobotia pappenheimi   123

Gobiobotia tungi   123

Gobiobotinae   123

gobiobotine   151

Gobiocypris   29, 149, 151

Gobiocypris rarus   6, 34

Gobioninae   4, 9, 126, 150, 151

gobionine   9, 147, 149, 151

Gymnocypris   73, 150

Gymnocypris chilianensis   74

Gymnocypris dobula   73

Gymnocypris eckloni   74

Gymnocypris przewalskii   74

Gymnocypris waddelli   74

Gymnodiptychus   73, 150

Gymnodiptychus pachycheilus   74

H

Haludaria fasciatus   49

Hampala   149

Hampala dispar   45

Hampala macrolepidota   45

Hemibarbus   126, 151

Hemibarbus barbus   4, 126

Hemibarbus labeo   126, 150

Hemibarbus longirostris   126

Hemibarbus maculatus   126

Hemiculterella sauvagei   81

Hemiculterella wui   81

Hemiculter leucisculus   82

Hemiculter lucidus   82

Hemigrammocypris rasborella   33

Henicorhynchus siamensis   114

Hybognathus hankinsoni   101

Hypophthalmichthyinae   91, 147, 148

hypophthalmichthyine   9, 151

Hypophthalmichthys   147

Hypophthalmichthys molitrix   91, 148

Hypophthalmichthys nobilis   9, 17, 25, 91, 147

Hypsibarbus malcolmi   49

I

Ischikauia steenackeri   82, 149

K

Kottelatia brittani   29

L

Labeo boga   113

Labeo chrysophekadion   113

Labeo cylindricus   113

Labeo erythropterus   113

Labeo rohita   114

Labeo rosae   114

Labeoninae   113

labeonine   9, 147, 149, 151, 152

Labiobarbus festivus   115

Labiobarbus leptocheilus   116

Labiobarbus lineatus   116

Labiobarbus siamensis   116

Ladislavia taczanowski   128

Laubuca caeruleostigmata   32

Leptobarbus hoevenii   49

Leuciscinae   4, 100, 151

leuciscine   9, 147, 151

Leuciscus   9

Leuciscus baicalensis   100

Leuciscus idus   100

Leuciscus waleckii   100

Lobocheilos melanotaenia   114

Lobocheilos rhabdoura   114

Luciocypris langsoni   49

Luciosoma bleekeri   29

Luciosoma setigerum   29

Luxilus cornutus   101

Lythrurus umbratilis   101

M

Margariscus margarita   102

Megalobrama amblycephala   17, 25, 82

Megalobrama terminalis   82

Mesobola brevianalis   32

Metzia lineata   82

Microphysogobio amurensis   130

Microphysogobio brevirostris   130

Microphysogobio labeoides   130

Microphysogobio tungtingensis   130

Microrasbora kubotai   29

Microrasbora rubescens   29

Misgurnus anguillicaudatus   7

Mylopharyngodon   100, 147, 149, 151

Mylopharyngodon piceus   9, 21, 25, 103, 149

Mystacoleucus atridorsalis   49

Mystacoleucus chilopterus   49

Mystacoleucus lepturus   49

Mystacoleucus marginatus   49

N

Neolissochilus benasi   50

Neolissochilus hexagonolepis   50

Nipponocypris sieboldii   33

Nipponocypris temminckii   33

Nocomis biguttatus   103

Nocomis micropogon   103

Notemigonus crysoleucas   103

Notropis atherinoides   103

Notropis heterodon   103

Notropis heterolepis   103

Notropis rubellus   103

Notropis volucellus   103

O

Ochetobius   100, 151

Ochetobius elongatus   100

Onychostoma angustistomata   50

Onychostoma barbatulum   50

Onychostoma barbatum   50

Onychostoma elongatum   50

Onychostoma gerlachi   50

Onychostoma leptura   50

Onychostoma lini   50

Onychostoma macrolepis   51

Onychostoma ovale   51

Onychostoma rarum   51

Onychostoma simum   51

Opsariichthys bidens   4, 15, 25, 32

Opsariichthys pachycephalus   32

Opsariichthys uncirostris   4, 32

Opsarius koratensis   29

Oreichthys cosuatis   51

Osteobrama cotio   51

Osteochilus lini   115

Osteochilus melanopleurus   115

Osteochilus microcephalus   115

Osteochilus salsburyi   115

Osteochilus spilurus   115

Osteochilus vittatus   115

Osteochilus waandersii   115

Oxygaster anomalura   32

P

Parabramis pekinensis   82

Paracanthobrama guichenoti   127

Parachela hypophthalmus   32

Parachela siamensis   32

Parachela williaminae   32

Paralaubuca barroni   80

Paralaubuca riveroi   80

Paralaubuca typus   80

Pararasbora   29, 149

Pararasbora moltrechti   33

Parator zonatus   55

Parazacco   29, 149

Parazacco spilurus   33

Percocypris   149

Percocypris pingi   45

Percocypris regani   45

Phoxinus neogaeus   100

Phoxinus phoxinus   100

Phoxinus steindachneri   100

Pimephales notatus   101

Pimephales promelas   101

Plagiognathops   89, 150

Plagiognathops microlepis   89

Platysmacheilus exiguous   130

Poropuntius chonglingchungi   51

Poropuntius daliensis   51

Poropuntius deauratus   51

Poropuntius huangchuchieni   51

Poropuntius ikedai   52

Poropuntius krempfi   52

Poropuntius opisthopterus   52

Procypris   139, 151

Procypris rabaudi   139

Pseudobrama   89

Pseudobrama simoni   89

Pseudogobio esocinus   4, 128, 150

Pseudogobio vaillanti   128

Pseudorasbora elongata   129

Pseudorasbora parva   4, 21, 25, 129

Pseudorasbora pumila pumila   129

Pseudorasbora pumila subsp. A   129

Psudobrama   89, 150

Ptychobarbus   73, 150

Ptychobarbus chungtienensis   74

Ptychobarbus kaznakovi   74

Pungtungia herzi   4, 130

Puntioplithes   139, 151

Puntioplites bulu   139

Puntioplites facifer   139

Puntioplites proctozystron   139

Puntioplites waandersi   139

Puntius arulius   53

Puntius aurotaeniatus   53

Puntius bimaculatus   55

Puntius binotatus   53

Puntius brevis   55

Puntius chola   54

Puntius conchonius   53

Puntius gelius   54

Puntius lateristriga   53

Puntius nigrofasciatus   54

Puntius oligolepis   54

Puntius partipentazona   54

Puntius rhomboocellatus   54

Puntius sealei   54

Puntius sophore   54

Puntius tetrazona   54

Puntius titteya   54

Puntius vittatus   55

R

Raiamas guttatus   29

Rasbora argyrotaenia   31

Rasbora aurotaenia   30

Rasbora borapetensis   30

Rasbora caudimaculata   30

Rasbora cephalotaenia   30

Rasbora daniconius   30

Rasbora dusonensis   30

Rasbora einthovenii   30

Rasbora elegans   31

Rasbora reticulata   31

Rasbora sumatrana   30

Rasbora tornieri   31

Rasbora trilineata   30

Rasboroides vaterifloris   30

Rectoris luxiensis   115

Rhinichthys atratulus   101

Rhinichthys cataractae   101

Rhinogobio ventralis   127

Rhodeus amarus   93

Rhodeus atremius   92

Rhodeus fangi   92

Rhodeus ocellatus   18, 25, 92

Rhodeus sinensis   93

Rhodeus suigensis   92

Rhynchocypris percnurus   100

Rutilus rutilus   101

S

Salmophasia bacaila   30

Sarcocheilichthys biwaensis   129

Sarcocheilichthys davidi   128

Sarcocheilichthys parvus   128

Sarcocheilichthys sinensis   21, 25, 128

Sarcocheilichthys variegatus microoculis   4, 129

Sarcocheilichthys variegatus variegatus   129

Saurogobio immaculatus   129

Scardinius erythrophthalmus   103

Schizopygopsis   73, 150

Schizopygopsis malacanthus   74

Schizopygopsis pylzovi   74

Schizopygopsis stoliczkai   75

Schizopygopsis younghasbandi   75

Schizorhoracinae   73, 147

schizothoracine   147, 149–151

Index   163

Schizothorax   73, 150

Schizothorax dolichonema   73

Schizothorax grahami   73

Schizothorax labiatus   73

Schizothorax lissolabiatus   73

Schizothorax meridionalis   73

Schizothorax plagiostomus   73

Schizothorax prenanti   73

Schizothorax wangchiachii   73

Semotilus corporalis   102

Sinibrama macrops   82

Sinibrama melrosei   83

Sinocyclocheilus grahami   52

Sinocyclocheilus tingi   52

Sinocyclocheilus yangzongensis   52

Spinibarbus denticulatus   52

Spinibarbus holandi   52

Spinibarbus sinensis   15, 25, 52

Squalidus chankaensis biwae   128

Squalidus chankaensis chankanensis   127

Squalidus gracilis gracilis   128

Squalidus japonicus japonicus   128

Squaliobarbus   100, 151

Squaliobarbus curriculus   102

Systomus orphoides   52

T

Tanakia himantegus   92

Tanakia laceolata   92

Tanakia limbata   92

Tanakia tanago   92

Tinca   100, 151

Tinca tinca   21, 25, 103

Tor douronensis   53

Tor putitora   53

Tor sinensis   53

Tor soro   53

Tribolodon brandti   102

Tribolodon hakonensis   4-6, 9, 20, 25, 102, 147

Tribolodon nakamurai   102

Tribolodon sachalinensis   102

Trigonopoma pauciperforatum   30

X

Xenocypridinae   89, 100, 147, 151

xenocypridine   147, 150, 151

Xenocypris   89, 150

Xenocypris davidi   89

Xenocypris macrolepis   89

Xenophysogobio boulengeri   123

Z

Zacco platypus   4, 33

Subject

A

A row   1

adult dentition   1, 2, 6, 152

angle of grinding surface   153

anterior margin   147, 153

anterior position   152

anterior tooth   2, 4, 5, 7-9, 152

A-type larval dentition   5

B

B row   1

B-type larval dentition   5

bulge   29, 147, 149, 150, 153

C

C row   1

central position   152

central tooth   2-5, 7-9, 152

cervical line   152, 153

cobitid dentition   6

conical tooth   147

crown   152, 153

crown's shoulder   152

C-type larval dentition   5

cyprinid dentition   6-9

cyprinine tooth   9

D

dental formula   2, 4, 27

denticle   92, 152, 153

D-type larval dentition   5

E

enameloid layer   152

F

family progenitor   7, 152

functioning tooth   2

G

germ of replacement tooth   2

grinding surface   147, 152, 153

groove   147, 149, 150, 153

H

hamulus groove   80, 92, 113, 150–153

I

imperfect type of larval dentition   5, 153

J

jaw tooth   3

L

larval dentition   1–3, 6, 152

larval tooth   3, 6

lateral margin   147, 153

Leuciscus Stage   9

M

major row   1, 5, 6, 152

major-row tooth   6, 152

marginal ridge   147, 153

margin of grinding surface   153

medial margin   147, 153

minor row   1, 6, 7, 152

minor-row tooth   5, 6, 152

O

occlusal margin   152

P

perfect type of larval dentition   5, 152

pharyngeal bone   1, 2, 4, 7, 8

pharyngeal cavity   152

pharyngeal dentition   1, 2

pharyngeal tooth   1, 2, 9

polyphyodont   2, 3

posterior margin   147, 153

posterior position   152

posterior tooth   7, 8, 153

progenitor   1

R

replacement tooth   1, 2, 6, 7

replacement wave   1, 2, 6, 7, 153

ridge   29, 147, 149, 150, 153

S

secondary grinding surface   89, 92, 113, 151, 153

self-generating model   3

serrate margin   152, 153

sirration   152

shoulder of crown   147, 152, 153

Stage 1   16, 25, 147

Stage 2   16, 25, 147

Stage 3   16, 25, 147

Stage 4   16, 25, 147

Stage 5   16, 25, 147

stem progenitor   7, 153

subtempral fossa   8

T

tooth axis   147, 154

tooth crown   147, 153, 154

tooth family   1, 2, 6, 153

tooth foot   152, 154

tooth germ   1, 6, 7

tooth hook   147, 151, 152, 154

tooth name   153

tooth neck   152, 154

tooth position   1, 7, 153

tooth replacement   2

tooth row   153

transitional dentition   2, 153

Type 1   28

Type 2   28

Type 3   28

Type 4   28

Type 5   28

Type-1 tooth   27

Type-2 tooth   27

Type-3 tooth   27

Type-4 tooth   27

Type-5 tooth   27

U

U-shaped groove   147, 154

Z

Zahnreihe model   3